Encyclopaedia of

DNA Recombinant Technology

Encyclopaedia of DNA Recombinant Technology

Volume 2

Editor:
R.N. Kumar

AGROTECH PRESS
JAIPUR ▪ NEW DELHI

AGROTECH PRESS

Regd. Off. "Bony Residency", Gate No. 2
Opp. Tilak Public School, Vishweswariya Nagar,
Gopalpura Road, Jaipur - 302018, India
Ph: 0141 - 2761280, Telefax: 2761381
E-mail : agrotechpress@gmail.com

Delhi Office: 102, 1st Floor, Satyam House,
4327/3, Ansari Road, Darya Ganj,
New Delhi - 110002, India
Ph: 011-45652440
visit us: www.agrotechpress.com

Distributed By
KEYA INTERNATIONAL PRESS LLP
E-mail: keyainternationalpressllp@gmail.com

Published in India, 2024

ISBN : 9789384568214 (3 Vols. Set)

Typeset by Thomson Press

Printed and bound in India by Thomson Press Ltd.,
New Delhi

Contents

Preface

Molecular biology is the branch of biology that deals with the molecular basis of biological activity. This field overlaps with other areas of biology and chemistry, particularly genetics and biochemistry. Molecular biology chiefly concerns itself with understanding the interactions between the various systems of a cell, including the interactions between the different types of DNA, RNA and protein biosynthesis as well as learning how these interactions are regulated. Researchers in molecular biology use specific techniques native to molecular biology but increasingly combine these with techniques and ideas from genetics and biochemistry.

Recombinant DNA molecules are DNA sequences that result from the use of laboratory methods to bring together genetic material from multiple sources, creating sequences that would not otherwise be found in biological organisms. Recombinant DNA is possible because DNA molecules from all organisms share the same chemical structure; they differ only in the sequence of nucleotides within that identical overall structure. Consequently, when DNA from a foreign source is linked to host sequences that can drive DNA replication and then introduced into a host organism, the foreign DNA is replicated along with the host DNA. Molecular cloning is the laboratory process used to create recombinant DNA. It is one of two widely-used methods used to direct the replication of any specific DNA sequence chosen by the experimentalist. The fundamental difference between the two methods is that molecular cloning involves replication of the DNA within a living cell, while PCR replicates DNA in the test tube, free of living cells. Formation of recombinant DNA requires a cloning vector, a DNA molecule that will replicate within a living cell. Vectors are generally derived from plasmids or viruses, and represent relatively small segments of DNA that contain necessary genetic signals for replication, as well as additional elements for convenience in inserting foreign DNA, identifying cells that contain recombinant DNA, and, where appropriate, expressing the foreign DNA. The choice of vector for molecular cloning depends on the choice of host organism, the size of the DNA to be cloned, and whether and how the foreign

DNA is to be expressed. The DNA segments can be combined by using a variety of methods, such as restriction enzyme/ligase cloning or Gibson assembly. Recombinant DNA is widely used in biotechnology, medicine and research. Today, recombinant proteins and other products that result from the use of rDNA technology are found in essentially every western pharmacy, doctor's or veterinarian's office, medical testing laboratory, and biological research laboratory. In addition, organisms that have been manipulated using recombinant DNA technology, and products derived from those organisms have found their way into many farms, supermarkets, home medicine cabinets and even pet shops. The most common application of recombinant DNA is in basic research, where it is important to most current work in the biological and biomedical sciences. Recombinant DNA is used to identify, map and sequence genes, and to determine their function. rDNA probes are employed in analysing gene expression within individual cells, and throughout the tissues of whole organisms. Recombinant proteins are widely used as reagents in laboratory experiments and to generate antibody probes for examining protein synthesis within cells and organisms. Many additional practical applications of recombinant DNA are found in human and veterinary medicine, in agriculture, and in bioengineering. Each of the three widely-used methods for diagnosing HIV infection has been developed using recombinant DNA. The antibody test uses a recombinant HIV protein to test for the presence of antibodies that the body has produced in response to an HIV infection. The DNA test looks for the presence of HIV genetic material using reverse transcriptase polymerase chain reaction. Development of the RT-PCR test was made possible by the molecular cloning and sequence analysis of HIV genomes.

The purpose of this book is to provide an insight on theoretical aspects of the subject. The book has been designed for under-graduates, post-graduates and technicians who wish to know and use the principles and techniques of Recombinant DNA Technology.

—Editor

1

Molecular Biotechnology

Molecular biotechnology is an exciting revolutionary scientific discipline that is based on the ability of a researcher to transfer specific units of genetic information from one organism to another. The objective of recombinant DNA technology is often to produce a useful product or a commercial process. In early 1970, traditional biotechnology was not well known as a scientific discipline. Research in this area was carried out in chemical engineering departments.

The term "biotechnology" was created in 1919 by a Hungarian engineer, Karl Ereky. According to Ereky, "biotechnology involves all works carried out with the aid of living things." More formally, biotechnology may be defined as "the application of scientific and engineering principles to the processing of material by biological agents to provide goods and services."

In 1961, a Swedish microbiologist Carl Gordon Heden redefined biotechnology as "the industrial production of goods and services by processing using biological organisms, and it is firmly ground on expertise in microbiology, biochemistry and chemical engineering." However the nature of biotechnology was changed forever by the development of recombinant DNA technology. With these techniques, the optimization of any biotechnological process was achieved more directly. Genetic engineering provided a means to create, rather than merely isolate highly productive strains.

Microorganisms and eukaryotic cells could be used as "biological factories" for the production of insulin, interferon, growth hormone, viral antigens and a variety of other proteins. Recombinant DNA technology could also be used to facilitate the biological production of large amounts of useful lower molecular weight compounds and macromolecules that occur naturally in minute quantities. This, technology facilitated the development of radically new medical therapies and diagnostic systems. Thus, the union of recombinant DNA technology with biotechnology created a vibrant, highly competitive field of study that has been called "Molecular Biotechnology." Molecular biotechnology ought to contribute unprecedented benefits to humanity.

1. It should provide opportunities to accurately diagnose and prevent or cure a wide range of infectious and genetic diseases.
2. Significant increase in crop yield may be obtained by generating disease-pathogen- or herbicide-resistant varieties.
3. Microorganisms that will produce chemicals, antibiotics, polymers, amino acids, enzymes, etc. can be developed.
4. Livestock and other animals that have enhanced genetically determined attributes can be developed. Molecular biotechnology with much fuss and fanfare has become a comprehensive scientific venture, both commercially and academically, in a remarkably short time.
5. A number of new scientific and business publications are devoted to molecular biotechnology. Both graduate and undergraduate programmes and courses have been created at many universities throughout the world.

Although it is exciting and important to emphasize the positive aspects of new advances, there are also social concerns and consequences that must be addressed. Because molecular biotechnology is so broadly based, its potential impact on society must be considered.

Many issues have been considered and discussed extensively by government commissions, and debated by individuals in popular and academic publications. There has been an active and extensive participation by both scientists and the general public in deciding how molecular biotechnology should proceed, although some controversies still remain.

Finally, molecular biotechnology is the third scientific revolution after the industrial and computer revolution, which is going to change the lives and future of humankind on this earth. The ability to manipulate genetic material to achieve specified outcome in living organisms promises major changes in many aspects of modem life

The Technology that may Impact Our Life in Future!

"Distinct from traditional biotechnology, the modern methods require knowledge of and manipulation of genes, the functional units of inheritance, and the discipline that is concerned with the manipulation of genes for the purpose of producing useful goods and services using living organisms is known as molecular biotechnology. " — Molecular Biotechnology, 4th Edition, American Society for Microbiology

The Many Definitions of Molecular Biotechnology

- Molecular Biotechnology is a new research field that emerged as an outcome of recombinant DNA technology and traditional industrial microbiology.
- Molecular Biotechnology is an exciting field in science that is based on the ability of researchers to transfer specific units of genetic information from one organism to another, which relies on the technique of recombinant DNA technology (genetic engineering).
- Molecular Biotechnology is the use of laboratory techniques to study and find ways to modify nucleic acids and proteins for its applications in the areas such as medicine, animal health, agriculture, environment etc.
- Molecular Biotechnology is the result of fusion of areas of research, such as molecular biology, microbiology, biochemistry, immunology, genetics, and cell biology. It is an exciting field fuelled by the ability to transfer genetic information between organisms with the goal of understanding important biological processes or creating a useful product which benefits mankind.

DNA

Deoxyribonucleic acid or DNA, is a nucleic acid that contains the genetic instructions used in the development and functioning of all known living organisms (with the exception of RNA viruses).

The main role of DNA molecules is the long-term storage of information. DNA is often compared to a set of blueprints, like a recipe or a code, since it contains the instructions needed to construct other components of cells, such as proteins and RNA molecules. The DNA segments that carry this genetic information are called genes, but other DNA sequences have structural purposes, or are involved in regulating the use of this genetic information.

DNA consists of two long polymers of simple units called nucleotides, with backbones made of sugars and phosphate groups joined by ester bonds. These two strands run in opposite directions to each other and are therefore anti-parallel. Attached to each sugar is one of four types of molecules called bases. It is the sequence of these four bases along the backbone that encodes information. This information is read using the genetic code, which specifies the sequence of the amino acids within proteins. The code is read by copying stretches of DNA into the related nucleic acid RNA, in a process called transcription. Within cells, DNA is organized into long structures called chromosomes. These chromosomes are duplicated before cells divide, in a process called DNA replication. Eukaryotic organisms (animals, plants, fungi, and protists) store most of their DNA inside the cell nucleus and some of their DNA in organelles, such as mitochondria or chloroplasts.

In contrast, prokaryotes (bacteria and archaea) store their DNA only in the cytoplasm. Within the chromosomes, chromatin proteins such as histones compact and organize DNA. These compact structures guide the interactions between DNA and other proteins, helping control which parts of the DNA are transcribed.

Properties

DNA is a long polymer made from repeating units called nucleotides. As first discovered by James D. Watson and Francis Crick, the structure of DNA of all species comprises two helical chains each coiled round the same axis, and each with a pitch of 34 Angstroms (3.4 nanometres) and a radius of 10 Angstroms (1.0 nanometres). According to another study, when measured in a particular solution, the DNA chain measured 22 to 26 Angstroms wide (2.2 to 2.6 nanometres), and one nucleotide unit measured 3.3 Å (0.33 nm) long. Although each individual repeating unit is very small, DNA polymers can be very large molecules containing millions of nucleotides. For instance, the largest human chromosome, chromosome number 1, is approximately 220 million base pairs long.

In living organisms, DNA does not usually exist as a single molecule, but instead as a pair of molecules that are held tightly together. These two long strands entwine like vines, in the shape of a double helix. The nucleotide repeats contain both the segment of the backbone of the molecule, which holds the chain together, and a base, which interacts with the other DNA strand in the helix. A base linked to a sugar is called a nucleoside and a base linked to a sugar and one or more phosphate groups is called a nucleotide. If multiple nucleotides are linked together, as in DNA, this polymer is called a polynucleotide.

The backbone of the DNA strand is made from alternating phosphate and sugar residues. The sugar in DNA is 2-deoxyribose, which is a pentose (five-carbon) sugar. The sugars are joined together by phosphate groups that form phosphodiester bonds between the third and fifth carbon atoms of adjacent sugar rings. These asymmetric bonds mean a strand of DNA has a direction. In a double helix the direction of the nucleotides in one strand is opposite to their direction in the other strand: the strands are *antiparallel*. The asymmetric ends of DNA strands are called the 52 (*five prime*) and 32 (*three prime*) ends, with the 5' end having a terminal phosphate group and the 3' end a terminal hydroxyl group. One major difference between DNA and RNA is the sugar, with the 2-deoxyribose in DNA being replaced by the alternative pentose sugar ribose in RNA.

The DNA double helix is stabilized by hydrogen bonds between the bases attached to the two strands. The four bases found in DNA are adenine (abbreviated A), cytosine (C), guanine (G) and thymine (T). These four bases are attached to the sugar/phosphate to form the complete nucleotide, as shown for adenosine monophosphate.

These bases are classified into two types; adenine and guanine are fused five- and six-membered heterocyclic compounds called purines, while cytosine and thymine are six-membered rings called pyrimidines. A fifth pyrimidine base, called uracil (U), usually takes the place of thymine in RNA and differs from thymine by lacking a methyl group on its ring. Uracil is not usually found in DNA, occurring only as a breakdown product of cytosine. In addition to RNA and DNA, a large number of artificial nucleic acid analogues have also been created to study the proprieties of nucleic acids, or for use in biotechnology.

Grooves

Twin helical strands form the DNA backbone. Another double helix may be found by tracing the spaces, or grooves, between the strands. These voids are adjacent to the base pairs and may provide a binding site. As the strands are not directly opposite each other, the grooves are unequally sized. One groove, the major groove, is 22 Å wide and the other, the minor groove, is 12 Å wide.

The narrowness of the minor groove means that the edges of the bases are more accessible in the major groove. As a result, proteins like transcription factors that can bind to specific sequences in double-stranded DNA usually make contacts to the sides of the bases exposed in the major groove. This situation varies in unusual conformations of DNA within the cell, but the major and minor grooves are always named to reflect the differences in size that would be seen if the DNA is twisted back into the ordinary B form.

Base Pair

In molecular biology, two nucleotides on opposite complementary DNA or RNA strands that are connected via hydrogen bonds are called a base pair (often abbreviated bp). In the canonical Watson-Crick DNA base pairing, adenine (A) forms a base pair with thymine (T) and guanine (G) forms a base pair with cytosine (C). In RNA, thymine is replaced by uracil (U). Alternate hydrogen bonding patterns, such as the wobble base pair and Hoogsteen base pair, also occur—particularly in RNA—giving rise to complex and functional tertiary structures. Importantly, pairing is the mechanism by which codons on messenger RNA molecules are recognized by anticodons on transfer RNA during protein translation. Some DNA- or RNA-binding enzymes can recognize specific base pairing patterns that identify particular regulatory regions of genes.

The size of an individual gene or an organism's entire genome is often measured in base pairs because DNA is usually double-stranded. Hence, the number of total base pairs is equal to the number of nucleotides in one of the strands (with the exception of non-coding single-stranded regions of telomeres). The haploid human genome (23 chromosomes) is estimated to be about 3 billion base pairs long and to contain 20,000–25,000 distinct genes. A kilobase (kb) is a unit of measurement in molecular biology equal to 1000 base pairs of DNA or RNA.

Hydrogen Bonding and Stability

Hydrogen bonding is the chemical interaction that underlies the base-pairing rules described above. Appropriate geometrical correspondence of hydrogen bond donors and acceptors allows only the "right" pairs to form stably. DNA with high GC-content is more stable than DNA with low GC-content, but contrary to popular belief, the hydrogen bonds do not stabilize the DNA significantly and stabilization is mainly due to stacking interactions.

The larger nucleobases, adenine and guanine, are members of a class of doubly-ringed chemical structures called purines; the smaller nucleobases, cytosine and thymine (and uracil), are members of a class of singly-ringed chemical structures called pyrimidines. Purines are only complementary with pyrimidines: pyrimidine-pyrimidine pairings are energetically unfavourable because the molecules are too far apart for hydrogen bonding to be established; purine-purine pairings are energetically unfavourable because the molecules are too close, leading to overlap repulsion. The only other possible pairings are GT and AC; these pairings are mismatches because the pattern of hydrogen donors and acceptors do not correspond. The GU pairing, with two hydrogen bonds, does occur fairly often in RNA.

Paired DNA and RNA molecules are comparatively stable at room temperature but the two nucleotide strands will separate above a melting point that is determined by the length of the molecules, the extent of mispairing (if any), and the GC content. Higher GC content results in higher melting temperatures; it is therefore unsurprising that the genomes of extremophile organisms such as *Thermus thermophilus* are particularly GC-rich. Conversely, regions of a genome that need to separate frequently—for example, the promoter regions for often-transcribed genes—are comparatively GC-poor. GC content and melting temperature must also be taken into account when designing primers for PCR reactions.

Base Stacking

Base stacking interactions in DNA and RNA are due to dispersion attraction, short-range exchange repulsion, and electrostatic interactions which also contribute to stability . Again GC stacking interactions with adjacent bases tend to be more favourable. (Note, though, that a GC stacking interaction with the next base pair is geometrically different from a CG interaction.) Base stacking effects are especially important in the secondary structure and tertiary structure of RNA; for example, RNA stem-loop structures are stabilized by base stacking in the loop region.

Base Analogs and Intercalators

Chemical analogs of nucleotides can take the place of proper nucleotides and establish non-canonical base-pairing, leading to errors (mostly point mutations) in DNA replication and DNA transcription. One common mutagenic base analog is 5-bromouracil, which resembles thymine but can base-pair to guanine in its enol form.

Other chemicals, known as DNA intercalators, fit into the gap between adjacent bases on a single strand and induce frameshift mutations by "masquerading" as a base, causing the DNA replication machinery to skip or insert additional nucleotides at the intercalated site. Most intercalators are large polyaromatic compounds and are known or suspected carcinogens. Examples include ethidium bromide and acridine.

Examples

The following DNA sequences illustrate pair double-stranded patterns. By convention, the top strand is written from the 5' end to the 3' end; thus the bottom strand is written 3' to 5'.

A base-paired DNA sequence:

ATCGATTGAGCTCTAGCG

TAGCTAACTCGAGATCGC

The corresponding RNA sequence, in which uracil is substituted for thymine where uracil takes its place in the RNA strand:

AUCGAUUGAGCUCUAGCG

UAGCUAACUCGAGAUCGC

Length Measurements

The following abbreviations are commonly used to describe the length of a D/RNA molecule:

- bp = base pair(s)—one bp corresponds to circa 3.4 Å of length along the strand

- kb (= kbp) = kilo base pairs = 1,000 bp
- Mb = mega base pairs = 1,000,000 bp
- Gb = giga base pairs = 1,000,000,000 bp.

In case of single stranded DNA/RNA units of nucleotides are used, abbreviated nt (or knt, Mnt, Gnt), as they are not paired. For distinction between units of computer storage and bases kbp, Mbp, Gbp etc. may be used for basepairs. The length of 16S rDNA for bacteria is 1542 base-pair in length.

The Centimorgan is also often used to imply distance along a chromosome, but the number of base-pairs it corresponds to varies widely. In the Human genome, it is about 1 million base pairs.

Sense (Molecular Biology)

Sense, when applied in a molecular biology context, is a general concept used to compare the polarity of nucleic acid molecules, such as DNA or RNA, to other nucleic acid molecules. Depending on the context within molecular biology, sense may have slightly different meanings.

DNA Sense

Molecular biologists call a single strand of DNA sense (or positive (+) sense) if an RNA version of the same sequence is translated or translatable into protein. Its complementary strand is called antisense (or negative (-) sense). Sometimes the phrase coding strand is encountered; however, protein coding and non-coding RNA's can be transcribed similarly from both strands, in some cases being transcribed in both directions from a common promoter region, or being transcribed from within introns, on both strands.

Antisense DNA

DNA normally has two strands, i.e., the sense strand and the antisense strand. In double-stranded DNA, only one strand codes for the RNA that is translated into protein. This DNA strand is referred to as the antisense strand. The strand that does not code for RNA is called the sense strand because it has a similar sequence to the messenger RNA (mRNA). Both the sense DNA strand and the mRNA transcript are complementary to the template DNA strand. Note that the DNA strands called “sense” and “antisense” are sometimes switched in older textbooks.

Antisense molecules interact with complementary strands of nucleic acids, modifying expression of genes.

Example with Double Stranded DNA :

DNA strand 1: antisense strand (copied to)→ RNA strand (sense).

DNA strand 2: sense strand.

Some regions within a double strand of DNA code for genes, which are usually instructions specifying the order of amino acids in a protein along with regulatory sequences, splicing sites, noncoding introns, and other complicating details. For a cell to use this information, one strand of the DNA serves as a template for the synthesis of a complementary strand of RNA.

The template DNA strand is called the transcribed strand with antisense sequence and the mRNA transcript is said to be sense sequence (the complement of antisense). Because the DNA is double-stranded, the strand complementary to the antisense sequence is called non-transcribed strand and has the same sense sequence as the mRNA transcript (though T bases in DNA are substituted with U bases in RNA).

A note on the confusion between "sense" and "antisense" strands: The strand names actually depend on which direction you are writing the sequence that contains the information for proteins (the "sense" information), not on which strand is on the top or bottom (that is arbitrary).

The only real biological information that is important for labeling strands is the location of the 5' phosphate group and the 3' hydroxyl group because these ends determine the direction of transcription and translation. A sequence 5' CGCTAT 3' is equivalent to a sequence written 3' TATCGC 5' as long as the 5' and 3' ends are noted.

If the ends are not labelled, convention is to assume that the sequence is written in the 5' to 3' direction. Good rule of thumb for figuring out the "sense" strand: Look for the start codon ATG (AUG in mRNA). In the table example, the sense mRNA has the AUG codon at the end (remember that translation proceeds in the 5' to 3' direction).

Ambisense

A single-stranded genome that contains both positive-sense and negative-sense is said to be ambisense. Bunya viruses have 3 single-stranded RNA (ssRNA) fragments containing both positive-sense and negative-sense sections; arenaviruses are also ssRNA viruses with an ambisense genome, as they have 2 fragments that are mainly negative-sense except for part of the 5' ends of the large and small segments of their genome.

Antisense RNA

Antisense RNA is a single-stranded RNA that is complementary to a messenger RNA (mRNA) strand transcribed within a cell. Antisense RNA may be introduced into a cell to inhibit translation of a complementary mRNA by base pairing to it and physically obstructing the translation machinery. This effect is therefore stoichiometric. An example of naturally occurring mRNA antisense mechanism is the hok/sok system of the E.coli R1 plasmid.

Antisense RNA has long been thought of as a promising technique for disease therapy; the only such case to have reached the market is the drug fomivirsen. One commentator has characterized antisense RNA as one of "dozens of technologies that are gorgeous in concept, but exasperating in [commercialization]". Generally, antisense RNA still lack effective design, biological activity, and efficient route of administration.

Historically, the effects of antisense RNA have often been confused with the effects of RNA interference (RNAi), a related process in which *double-stranded* RNA fragments called small interfering RNAs trigger catalytically mediated gene silencing, most typically by targeting the RNA-induced silencing complex (RISC) to bind to and degrade the mRNA. Attempts to genetically engineer transgenic plants to express antisense RNA instead activate the RNAi pathway, although the processes result in differing magnitudes of the same downstream effect, gene silencing. Well-known examples include the Flavr Savr tomato and two cultivars of ringspot-resistant papaya.

Transcription of longer cis-antisense transcripts is a common phenomenon in the mammalian transcriptome. Although the function of some cases have been described, such as the Zeb2/Sip1 antisense RNA, no general function has been elucidated. In the case of Zeb2/Sip1, the antisense noncoding RNA is opposite the 5' splice site of an intron in the 5'UTR of the Zeb2 mRNA.

Expression of the antisense ncRNA prevents splicing of an intron that contains a ribosome entry site necessary for efficient expression of the Zeb2 protein. Transcription of long antisense ncRNAs is often concordant with the associated protein-coding gene, but more detailed studies have revealed that the relative expression patterns of the mRNA and antisense ncRNA are complex.

RNA Sense in Viruses

In virology, the genome of an RNA virus can be said to be either positive-sense, also known as a "plus-strand", or negative-sense, also known as a "minus-strand". In most cases, the terms *sense* and *strand* are used interchangeably, making such terms as *positive-strand* equivalent to *positive-sense*, and *plus-strand* equivalent to *plus-sense*. Whether a virus genome is positive-sense or negative-sense can be used as a basis for classifying viruses.

Positive-sense

Positive-sense (5' to 3') viral RNA signifies that a particular viral RNA sequence may be directly translated into the desired viral proteins. Therefore, in positive-sense RNA viruses, the viral RNA genome can be considered viral mRNA, and can be immediately translated by the host cell. Unlike negative-sense RNA, positive-sense RNA is of the same sense as mRNA. Some viruses (e.g., Coronaviridae) have positive-sense genomes that can act as mRNA and be used directly to synthesise proteins without the help of a complementary RNA intermediate. Because of this, these viruses do not need to have an RNA polymerase packaged into the virion.

Negative-sense

Negative-sense (3' to 5') viral RNA is complementary to the viral mRNA and thus must be converted to positive-sense RNA by an RNA polymerase prior to translation. Negative-sense RNA (like DNA) has a nucleotide sequence complementary to the mRNA that it encodes. Like DNA, this RNA cannot be translated into protein directly. Instead, it must first be transcribed into a positive-sense RNA that acts as an mRNA. Some viruses (Influenza, for example) have negative-sense genomes and so must carry an RNA polymerase inside the virion.

Enzymes

Enzyme-dependent antisense includes forms dependent on RNase H activity to degrade target mRNA, including single-stranded DNA, RNA, and phosphorothioate antisense. The R1 plasmid hok/sok system is an example of mRNA antisense regulation process, through enzymatic degradation of the resulting RNA duplex. Double stranded RNA acts as enzyme-dependent antisense through the RNAi/siRNA pathway, involving target mRNA recognition through sense-antisense strand pairing followed by target mRNA degradation by the RNA-induced silencing complex (RISC).

Steric blocking antisense (RNase-H-independent antisense) interferes with gene expression or other mRNA-dependent cellular processes by binding to a target sequence of mRNA and getting in the way of other processes. Steric blocking antisense includes 2'-O alkyl (usually in chimeras with RNase-H dependent antisense), peptide nucleic acid (PNA), locked nucleic acid (LNA), and Morpholino antisense.

DNA Supercoil

In a "relaxed" double-helical segment of B-DNA, the two strands twist around the helical axis once every 10.4-10.5 base pairs of sequence. Adding or subtracting twists, as some enzymes can do, imposes strain. If a DNA segment under twist strain were to be closed into a circle by joining its two ends and then it is allowed to move freely, the circular DNA would contort into a new shape. Such a contortion is a supercoil. For each additional helical twist being accommodated, the lobes will show one more rotation about their axis.

The noun form "supercoil" is rarely used in the context of DNA topology. Instead, global contortions of a circular DNA, such as the rotation, are referred to as *writhe*. "Supercoiling" is an abstract mathematical property representing the sum of twist and writhe. The twist is the number of helical turns in the DNA and the writhe is the number of times the double helix crosses over on itself (these are the supercoils).

Extra helical twists are positive and lead to positive supercoiling, while subtractive twisting causes negative supercoiling. Many topoisomerase enzymes sense supercoiling and either generate or dissipate it as they change DNA topology. DNA of most organisms is negatively supercoiled.

In part because chromosomes may be very large, segments in the middle may act as if their ends are anchored. As a result, they may be unable to distribute excess twist to the rest of the chromosome or to absorb twist to recover from underwinding—the segments may become *supercoiled*, in other words. In response to supercoiling, they will assume an amount of writhe, just as if their ends were joined.

Supercoiled DNA forms two structures; a plectoneme or a toroid, or a combination of both. A negatively supercoiled DNA molecule will produce either a one-start left-handed helix, the toroid, or a two-start right-handed helix with terminal loops, the plectoneme. Plectonemes are typically more common in nature, and this is the shape most bacterial plasmids will take.

For larger molecules it is common for hybrid structures to form - a loop on a toroid can extend into a plectoneme. If all the loops on a toroid extend then it becomes a branch point in the plectonemic structure.

Occurrence of DNA Supercoiling

DNA supercoiling is important for DNA packaging within all cells. Because the length of DNA can be thousands of times that of a cell, packaging this genetic material into the cell or nucleus (in eukaryotes) is a difficult feat. Supercoiling of DNA reduces the space and allows for a lot more DNA to be packaged. In prokaryotes, plectonemic supercoils are predominant, because of the circular chromosome and relatively small amount of genetic material.

In eukaryotes, DNA supercoiling exists on many levels of both plectonemic and solenoidal supercoils, with the solenoidal supercoiling proving most effective in compacting the DNA. Solenoidal supercoiling is achieved with histones to form a 10nm fibre. This fibre is further coiled into a 30nm fibre, and further coiled upon itself numerous times more.

DNA packaging is greatly increased during nuclear division events such as mitosis or meiosis, where DNA must be compacted and segregated to daughter cells. Condensins and cohesins are *Structural Maintenance of Chromosome* proteins that aid in the condensation of sister chromatids and the linkage of the centromere in sister chromatids. These SMC proteins induce positive supercoils.

Supercoiling is also required for DNA/RNA synthesis. Because DNA must be unwound for DNA/RNA polymerase action, supercoils will result. The region ahead of the polymerase complex will be unwound; this stress is compensated with positive supercoils ahead of the complex. Behind the complex, DNA is rewound and there will be compensatory negative supercoils.

It is important to note that topoisomerases such as DNA gyrase (Type II Topoisomerase) play a role in relieving some of the stress during DNA/RNA synthesis.

Modelling using Mathematics

DNA supercoiling can be described numerically by changes in the 'linking number' Lk. The linking number is the most descriptive property of supercoiled DNA. Lk_o, the number of turns in the relaxed (B type) DNA plasmid/molecule, is determined by dividing the total base pairs of the molecule by the relaxed bp/turn which, depending on reference is 10.4-10.5.

$$Lk_o = bp / 10.4$$

Lk is merely the number of crosses a single strand makes across the other in a planar projection. The topology of the DNA is described by the equation below in which the linking number is equivalent to the sum of TW, which is the number of twists or turns of the double helix, and Wr which is the number of coils or 'writhes'.

If there is a closed DNA molecule, the sum of TW and Wr, or the linking number, does not change. However, there may be complementary changes in TW and Wr without changing their sum.

$$Lk = Tw + Wr$$

The change in the linking number, ΔLk, is the actual number of turns in the plasmid/molecule, Lk, minus the number of turns in the relaxed plasmid/ molecule Lk_o.

$$\Delta Lk = Lk - Lk_o$$

If the DNA is negatively supercoiled $\Delta Lk < 0$. The negative supercoiling implies that the DNA is underwound. A standard expression independent of the molecule size is the "specific linking difference" or "superhelical density" denoted σ. σ represents the number of turns added or removed relative to the total number of turns in the relaxed molecule/plasmid, indicating the level of supercoiling.

$$\sigma = \Delta Lk / Lk_o$$

The Gibbs free energy associated with the coiling is given by the equation below

$$\Delta G / N = 10RT\sigma^2$$

Examples

Since the linking number L of supercoiled DNA is the number of times the two strands are intertwined (and both strands remain covalently intact), L cannot change. The reference state (or parameter) L_0 of a circular DNA duplex is its relaxed state. In this state, its writhe $W = 0$. Since $L = T + W$, in a relaxed state $T = L$. Thus, if we have a 400 bp relaxed circular DNA duplex, $L \sim 40$ (assuming ~10 bp per turn in B-DNA). Then $T \sim 40$.

Positively supercoiling:

$$T = 0,\ W = 0,\ \text{then } L = 0$$
$$T = +3,\ W = 0,\ \text{then } L = +3$$
$$T = +2,\ W = +1,\ \text{then } L = +3$$

Negatively supercoiling:

$$T = 0,\ W = 0,\ \text{then } L = 0$$
$$T = -3,\ W = 0,\ \text{then } L = -3$$
$$T = -2,\ W = -1,\ \text{then } L = -3$$

Negative supercoils favor local unwinding of the DNA, allowing processes such as transcription, DNA replication, and recombination. Negative supercoiling is also thought to favour the transition between B-DNA and Z-DNA, and moderate the interactions of DNA binding proteins involved in gene regulation.

Molecular Structure of Nucleic Acids: A Structure for Deoxyribose Nucleic Acid

The *Molecular structure of Nucleic Acids: A Structure for Deoxyribose Nucleic Acid* was an article published by James D. Watson and Francis Crick in the scientific journal *Nature* in its 171st volume on page 737-738 (dated April 25, 1953). It was the first publication which described the discovery of the double helix structure of DNA. This discovery had a major impact on biology, particularly in the field of genetics.

This article is often termed a "pearl" of science because it is brief and contains the answer to a fundamental mystery about living organisms. This mystery was the question of how it was possible that genetic instructions were held inside organisms and how they were passed from generation to generation. The article presents a simple and elegant solution, which surprised many biologists at the time who felt that DNA transmission was going to be more difficult to detail and understand.

The Nature of the Discovery

Watson and Crick's 1953 article contains the answer to a fundamental mystery about living organisms. The nature of their discovery was distinctive and in some ways unsurprising. What is hidden in the technical jargon of the title is that Watson and Crick's discovery of the chemical structure of DNA finally revealed how genetic instructions are stored inside organisms and passed from generation to generation. However, Watson and Crick took the majority of their information from a little known scientist named Rosalind Franklin, without her knowledge. She discovered the exact structure of DNA before either of the two, however she died before the public could find out about her. Watson never and to this day, has not given her credit for the discovery.

Origins of Molecular Biology

The application of physics and chemistry to biological problems led to the development of molecular biology. Not all biology that concerns molecules falls into the category that is labelled "molecular biology". Molecular biology is particularly concerned with the flow and consequences of biological information at the level of genes and proteins. The discovery of the DNA double helix made clear that genes are functionally defined parts of DNA molecules and that there must be a way for cells to make use of their DNA genes in order to make proteins.

Linus Pauling was a chemist who was very influential in developing an understanding of the structure of biological molecules. In 1951, Pauling published the structure of the alpha helix, a fundamentally important structural component of proteins. Early in 1953 Pauling published an incorrect triple helix model of DNA. Both Crick, and particularly Watson, felt that they were racing against Pauling to discover the structure of DNA.

Max Delbruck was a physicist who recognized some of the biological implications of quantum physics. Delbruck's thinking about the physical basis of life stimulated Erwin Schrodinger to write the highly influential book, *What Is Life?* Schrodinger's book was an important influence on Francis Crick, James D. Watson, and Maurice Wilkins who won a Nobel prize in Medicine for the discovery of the DNA double helix. Delbruck's efforts to promote the "Phage Group" (exploring genetics by way of the viruses that infect bacteria) was important in the early development of molecular biology in general and the development of Watson's scientific interests in particular.

DNA Structure and Function

It is not always the case that the structure of a molecule is easy to relate to its function. What makes the structure of DNA so obviously related to its function was described modestly at the end of the article: "It has not escaped our notice that the specific pairing we have postulated immediately suggests a possible copying mechanism for the genetic material". The "specific pairing" is a key feature of the Watson and Crick model of DNA, the pairing of nucleotide subunits. In DNA, the amount of guanine is equal to cytosine and the amount of adenine is equal to thymine. The A:T and C:G pairs are structurally similar. In particular, the length of each base pair is the same and they fit equally between the two phosphate backbones. The base pairs are held together by hydrogen bonds, a type of chemical attraction that is easy to break and easy to reform. After realizing the structural similarity of the A:T and C:G pairs, Watson and Crick soon produced their double helix model of DNA with the hydrogen bonds at the core of the helix providing a way to unzip the two complementary strands for easy replication: the last key requirement for a likely model of the genetic molecule.

Indeed, the base-pairing did suggest a way to copy a DNA molecule. Just pull apart the two phosphate backbones, each with its hydrogen bonded A, T, G, and C components. Each strand could then be used as a template for assembly of a new base-pair complementary strand.

Future Considerations

When Watson and Crick produced their double helix model of DNA, it was known that most of the specialized features of the many different life forms on Earth are made possible by proteins. Structurally, proteins are long chains of amino acid subunits. In some way, the genetic molecule, DNA, had to contain instructions for how to make the thousands of proteins found in cells. From the DNA double helix model, it was clear that there must be some correspondence between the linear sequences of nucleotides in DNA molecules to the linear sequences of amino acids in proteins. The details of how sequences of DNA instruct cells to make specific proteins was worked out by molecular biologists during the period from 1953 to 1965. Francis Crick played an integral role in both the theory and analysis of the experiments that led to an improved understanding of the genetic code.

Consequences

Other advances in molecular biology stemming from the discovery of the DNA double helix eventually led to ways to sequence genes. James Watson played an important role in getting government funding for the Human Genome Project. The ability to sequence and manipulate DNA is now central to the biotechnology industry and modern medicine. The austere beauty of the structure and the practical implications of the DNA double helix combined to make *Molecular structure of Nucleic Acids; A Structure for Deoxyribose Nucleic Acid* one of the most prominent biology articles of the twentieth century.

Controversy

Watson and Crick based their molecular model of the DNA double helix on data that had been collected by researchers in several other laboratories. Watson and Crick were the first to put together all of the scattered fragments of information that were required to produce a successful molecular model of DNA. Much of the data that were used by Crick and Watson came from unpublished work by Maurice Wilkins, Rosalind Franklin, A.R. Stokes, and H.R. Wilson at King's College London in the University of London. Key data from Wilkins, Stokes, and Wilson, and, separately, by Franklin and Gosling, were published in two separate additional articles in the same issue of *Nature* with the article by Watson and Crick. The article by Watson and Crick did acknowledge that they had been "stimulated" by experimental results from the King's College researchers, and a similar acknowledgment was published by M. H. F. Wilkins, A.R. Stokes, and H. R. Wilson in the following three-page article.

In 1968, Watson published a highly controversial autobiographical account of the discovery of the double-helical, molecular structure of DNA called *The Double Helix*, and which was not accepted—at least publicly—either by Francis Crick or by M.H.F. Wilkins. Furthermore, Erwin Chargaff also printed a rather "unsympathetic review" of James D. Watson's booklet in the March 29, 1968 issue of *Science*. In his `autobiographical' booklet, Watson stated among other things that he and Crick had access to some of Franklin's data from a source that she was not aware of, and also that he had seen—without her permission—the B-DNA X-ray diffraction pattern obtained by Franklin and Gosling in May 1952 at King's in London. In particular, in late 1952, Dr. Franklin had submitted a progress report to the Medical Research Council, which was reviewed by Dr. Max Perutz, then at The Cavendish Laboratory of the University of Cambridge, UK. Watson and Crick also worked in the MRC-supported Cavendish Laboratory in Cambridge whereas Drs. Wilkins and Franklin were in the MRC supported laboratory at King's in London. Such MRC reports were not usually widely circulated, but Crick read a copy of Dr. Franklin's research summary in early 1953.

Max Perutz's justification for passing this information to both Crick and Watson was that the report contained information which Watson has previously heard in November 1951 when Dr. Franklin talked about her unpublished results with Raymond Gosling during a meeting arranged by Dr. M.H.F. Wilkins at King's College, following a request from Crick and Watson; this justification does not hold however for Crick who was not present at this November 1951 meeting, but who also was given access by Max Perutz to Franklin's MRC report data which prompted Crick and Watson to seek permission from Sir Lawrence Bragg—who was at the time the head of the Cavendish Laboratory in Cambridge—to publish in *Nature* their double-helix molecular model of DNA based on Franklin's and also Wilkins's data. Moreover, in November 1951 Watson had acquired—by his own admission—little training in X-ray crystallography, and therefore had not fully understood (again, according to his own admission, in "The Double Helix") what Dr. Franklin was saying about the structural symmetry of the DNA molecule. Crick, however, knowing the Fourier transforms of Bessel functions that represent the X-ray diffraction patterns of helical structures of atoms, correctly interpreted further one of Dr.Franklin's experimental findings as indicating that DNA was most likely to be a double helix with the two polynucleotide chains running in opposite directions.

Crick was thus in a unique position to make this interpretation because he had previously worked on the X-ray diffraction data for other large molecules that had similar, helical symmetry to that of DNA. Dr. Franklin, on the other hand, rejected at first the molecular model building approach proposed by Crick and Watson because their first DNA model presented by Watson to her and Dr. M.H.F. Wilkins in 1952 in London had an obviously incorrect structure with hydrated charged groups on the inside of the model, rather than on the outside, as explicitly admitted by James D. Watson in his "Double Helix" booklet. It is therefore questioned whether Crick's colleague, Dr.Max Perutz, acted unethically by allowing Crick access to Dr. Franklin's MRC report about the crystallographic unit of the B-DNA and A-DNA structures. Dr. Perutz claimed, however, that he felt he had not because this report was not confidential, and had been designed as part of an effort to promote contact between different MRC research groups.

Molecular Models of DNA

Molecular models of DNA structures are representations of the molecular geometry and topology of Deoxyribonucleic acid (DNA) molecules using one of several means, with the aim of simplifying and presenting the essential, physical and chemical, properties of DNA molecular structures either *in vivo* or *in vitro*. These representations include closely packed spheres (CPK models) made of plastic, metal wires for 'skeletal models', graphic computations and animations by computers, artistic rendering. Computer molecular models also allow animations and molecular dynamics simulations that are very important for understanding how DNA functions *in vivo*.

The more advanced, computer-based molecular models of DNA involve molecular dynamics simulations as well as quantum mechanical computations of vibro-rotations, delocalized molecular orbitals (MOs), electric dipole moments, hydrogen-bonding, and so on. DNA molecular dynamics modelling involves simulations of DNA molecular geometry and topology changes with time as a result of both intra- and inter- molecular interactions of DNA. Whereas molecular models of Deoxyribonucleic acid (DNA) molecules such as closely packed spheres (CPK models) made of plastic or metal wires for 'skeletal models' are useful representations of static DNA structures, their usefulness is very limited for representing complex DNA dynamics. Computer molecular modelling allows both animations and molecular dynamics simulations that are very important for understanding how DNA functions *in vivo*.

History

History of Molecular Biology: The history of molecular biology begins in the 1930s with the convergence of various, previously distinct biological disciplines: biochemistry, genetics, microbiology, and virology. With the hope of understanding life at its most fundamental level, numerous physicists and chemists also took an interest in what would become molecular biology.

In its modern sense, molecular biology attempts to explain the phenomena of life starting from the macromolecular properties that generate them. Two categories of macromolecules in particular are the focus of the molecular biologist: 1) nucleic acids, among which the most famous is deoxyribonucleic acid (or DNA), the constituent of genes, and 2) proteins, which are the active agents of living organisms. One definition of the scope of molecular biology therefore is to characterize the structure, function and relationships between these two types of macromolecules. This relatively limited definition will suffice to allow us to establish a date for the so-called "molecular revolution", or at least to establish a chronology of its most fundamental developments.

General Overview

In its earliest manifestations, molecular biology—the name was coined by Warren Weaver of the Rockefeller Foundation in 1938—was an ideal of physical and chemical explanations of life, rather than a coherent discipline. Following the advent of the Mendelian-chromosome theory of heredity in the 1910s and the maturation of atomic theory and quantum mechanics in the 1920s, such explanations seemed within reach. Weaver and others encouraged (and funded) research at the intersection of biology, chemistry and physics, while prominent physicists such as Niels Bohr and Erwin Schrodinger turned their attention to biological speculation. However, in the 1930s and 1940s it was by no means clear which—if any—cross-disciplinary research would bear fruit; work in colloid chemistry, biophysics and radiation biology, crystallography, and other emerging fields all seemed promising.

In 1940, George Beadle and Edward Tatum demonstrated the existence of a precise relationship between genes and proteins. In the course of their experiments connecting genetics with biochemistry, they switched from the genetics mainstay *Drosophila* to a more appropriate model organism, the fungus *Neurospora*; the construction and exploitation of new model organisms would become a recurring theme in the development of molecular biology. In 1944, Oswald Avery, working at the Rockefeller Institute of New York, demonstrated that genes are made up of DNA. In 1952, Alfred Hershey and

Martha Chase confirmed that the genetic material of the bacteriophage, the virus which infects bacteria, is made up of DNA. In 1953, James Watson and Francis Crick discovered the double helical structure of the DNA molecule. In 1961, Francois Jacob and Jacques Monod hypothesized the existence of an intermediary between DNA and its protein products, which they called messenger RNA.

Between 1961 and 1965, the relationship between the information contained in DNA and the structure of proteins was determined: there is a code, the genetic code, which creates a correspondence between the succession of nucleotides in the DNA sequence and a series of amino acids in proteins. At the beginning of the 1960s, Monod and Jacob also demonstrated how certain specific proteins, called regulative proteins, latch onto DNA at the edges of the genes and control the transcription of these genes into messenger RNA; they direct the "expression" of the genes.

The chief discoveries of molecular biology took place in a period of only about twenty-five years. Another fifteen years were required before new and more sophisticated technologies, united today under the name of genetic engineering, would permit the isolation and characterization of genes, in particular those of highly complex organisms.

The Exploration of the Molecular Dominion

If we evaluate the molecular revolution within the context of biological history, it is easy to note that it is the culmination of a long process which began with the first observations through a microscope. The aim of these early researchers was to understand the functioning of living organisms by describing their organization at the microscopic level.

From the end of the 18th century, the characterization of the chemical molecules which make up living beings gained increasingly greater attention, along with the birth of physiological chemistry in the 19th century, developed by the German chemist Justus von Liebig and following the birth of biochemistry at the beginning of the 20th, thanks to another German chemist Eduard Buchner. Between the molecules studied by chemists and the tiny structures visible under the optical microscope, such as the cellular nucleus or the chromosomes, there was an obscure zone, "the world of the ignored dimensions," as it was called by the chemical-physicist Wolfgang Ostwald. This world is populated by colloids, chemical compounds whose structure and properties were not well defined. The successes of molecular biology derived from the exploration of that unknown world by means of the new technologies developed by chemists and physicists: X-ray diffraction, electron microscopy, ultracentrifugization, and electrophoresis. These studies revealed the structure and function of the macromolecules.

A milestone in that process was the work of Dr. Linus Pauling in 1949, which for the first time linked the specific genetic mutation in patients with sickle cell disease to a demonstrated change in an individual protein, the hemoglobin in the erythrocytes of heterozygous or homozygous individuals.

The Encounter between Biochemistry and Genetics

The development of molecular biology is also the encounter of two disciplines which made considerable progress in the course of the first thirty years of the twentieth century: biochemistry and genetics. The first studies the structure and function of the molecules which make up living things. Between 1900 and 1940, the central processes of metabolism were described: the process of digestion and the absorption of the nutritive elements derived from alimentation, such as the sugars. Every one of these processes is catalysed by a particular enzyme. Enzymes are proteins, like the antibodies present in blood or the proteins responsible for muscular contraction. As a consequence, the study of proteins, of their structure and synthesis, became one of the principal objectives of biochemists.

The second discipline of biology which developed at the beginning of the 20th century is genetics. After the rediscovery of the laws of Mendel through the studies of Hugo de Vries, Carl Correns and Erich von Tschermak in 1900, this science began to take shape thanks to the adoption by Thomas Hunt Morgan, in 1910, of a model organism for genetic studies, the famous fruit fly (Drosophila melanogaster).

Shortly after, Morgan showed that the genes are localized on chromosomes. Following this discovery, he continued working with Drosophila and, along with numerous other research groups, confirmed the importance of the gene in the life and development of organisms. Nevertheless, the chemical nature of genes and their mechanisms of action remained a mystery. Molecular biologists committed themselves to the determination of the structure, and the description of the complex relations between, genes and proteins.

The development of molecular biology was not just the fruit of some sort of intrinsic "necessity" in the history of ideas, but was a characteristically historical phenomenon, with all of its unknowns, imponderables and contingencies: the remarkable developments in physics at the beginning of the 20th century highlighted the relative lateness in development in biology, which became the "new frontier" in the search for knowledge about the empirical world. Moreover, the developments of the theory of information and cybernetics in the 1940s, in response to military exigencies, brought to the new biology a significant number of fertile ideas and, especially, metaphors.

The choice of bacteria and of its virus, the bacteriophage, as models for the study of the fundamental mechanisms of life was almost natural - they are the smallest living organisms known to exist - and at the same time the fruit of individual choices. This model owes its success, above all, to the fame and the sense of organization of Max Delbruck, a German physicist, who was able to create a dynamic research group, based in the United States, whose exclusive scope was the study of the bacteriophage: the *School of the Phage.*

The geographic panorama of the developments of the new biology was conditioned above all by preceding work. The US, where genetics had developed the most rapidly, and the UK, where there was a coexistence of both genetics and biochemical research of highly advanced levels, were in the avant-garde. Germany, the cradle of the revolutions in physics, with the best minds and the most advanced laboratories of genetics in the world, should have had a primary role in the development of molecular biology. But history decided differently: the arrival of the Nazis in 1933 - and, to a less extreme degree, the rigidification of totalitarian measures in fascist Italy - caused the emigration of a large number of Jewish and non-Jewish scientists. The majority of them fled to the US or the UK, providing an extra impulse to the scientific dynamism of those nations. These movements ultimately made molecular biology a truly international science from the very beginnings.

History of DNA Biochemistry

The study of DNA is a central part of molecular biology.

First Isolation of DNA

Working in the 19th century, biochemists initially isolated DNA and RNA (mixed together) from cell nuclei. They were relatively quick to appreciate the polymeric nature of their "nucleic acid" isolates, but realized only later that nucleotides were of two types—one containing ribose and the other deoxyribose. It was this subsequent discovery that led to the identification and naming of DNA as a substance distinct from RNA.

Friedrich Miescher (1844–1895) discovered a substance he called "nuclein" in 1869. Somewhat later, he isolated a pure sample of the material now known as DNA from the sperm of salmon, and in 1889 his pupil, Richard Altmann, named it "nucleic acid". This substance was found to exist only in the chromosomes. In 1919 Phoebus Levene at the Rockefeller Institute identified the components (the four bases, the sugar and the phosphate chain) and he showed that the components of DNA were linked in the order phosphate-sugar-base. He called each of these units a nucleotide and suggested the DNA molecule consisted of a string of nucleotide units linked together through the phosphate groups, which are the 'backbone' of the molecule. However Levene thought the chain was short and that the bases repeated in the same fixed order. Torbjorn Caspersson and Einar Hammersten showed that DNA was a polymer.

Chromosomes and Inherited Traits

Max Delbruck, Nikolai V. Timofeeff-Ressovsky, and Karl G. Zimmer published results in 1935 suggesting that chromosomes are very large molecules the structure of which can be changed by treatment with X-rays, and that by so changing their structure it was possible to change the heritable characteristics governed by those chromosomes. In 1937 William Astbury produced the first X-ray diffraction patterns from DNA. He was not able to propose the correct structure but the patterns showed that DNA had a regular structure and therefore it might be possible to deduce what this structure was.

In 1943, Oswald Theodore Avery and a team of scientists discovered that traits proper to the "smooth" form of the *Pneumococcus* could be transferred to the "rough" form of the same bacteria merely by making the killed "smooth" (S) form available to the live "rough" (R) form. Quite unexpectedly, the living R *Pneumococcus* bacteria were transformed into a new strain of the S form, and the transferred S characteristics turned out to be heritable. Avery called the medium of transfer of traits the transforming principle; he identified DNA as the transforming principle, and not protein as previously thought. He essentially redid Frederick Griffith's experiment. In 1953, Alfred Hershey and Martha Chase did an experiment (Hershey-Chase experiment) that showed, in T2 phage, that DNA is the genetic material (Hershey shared the Nobel prize with Luria).

Discovery of the Structure of DNA

In the 1950s, three groups made it their goal to determine the structure of DNA. The first group to start was at King's College London and was led by Maurice Wilkins and was later joined by Rosalind Franklin. Another group consisting of Francis Crick and James D. Watson was at Cambridge. A third group was at Caltech and was led by Linus Pauling. Crick and Watson built physical models using metal rods and balls, in which they incorporated the known chemical structures of the nucleotides, as well as the known position of the linkages joining one nucleotide to the next along the polymer. At King's College Maurice Wilkins and Rosalind Franklin examined X-ray diffraction patterns of DNA fibres. Of the three groups, only the London group was able to produce good quality diffraction patterns and thus produce sufficient quantitative data about the structure.

Helix Structure

In 1948 Pauling discovered that many proteins included helical shapes. Pauling had deduced this structure from X-ray patterns and from attempts to physically model the structures. (Pauling was also later to suggest an incorrect three chain helical DNA structure based on Astbury's data.) Even in the initial diffraction data from DNA by Maurice Wilkins, it was evident that the structure involved helices. But this insight was only a beginning. There remained the questions of how many strands came together, whether this number was the same for every helix, whether the bases pointed toward the helical axis or away, and ultimately what were the explicit angles and coordinates of all the bonds and atoms. Such questions motivated the modelling efforts of Watson and Crick.

Complementary Nucleotides

In their modelling, Watson and Crick restricted themselves to what they saw as chemically and biologically reasonable. Still, the breadth of possibilities was very wide. A breakthrough occurred in 1952, when Erwin Chargaff visited Cambridge and inspired Crick with a description of experiments Chargaff had published in 1947.

Chargaff had observed that the proportions of the four nucleotides vary between one DNA sample and the next, but that for particular pairs of nucleotides — adenine and thymine, guanine and cytosine — the two nucleotides are always present in equal proportions. Using X-ray diffraction, as well as other data from Rosalind Franklin and her information that the bases were paired, James D. Watson and Francis Crick arrived at the first accurate model of DNA's molecular structure in 1953, which was accepted through inspection by Rosalind Franklin.

The discovery was made on February 28, 1953; the first Watson/Crick paper appeared in Nature on April 25, 1953. Sir Lawrence Bragg, the director of the Cavendish Laboratory, where Watson and Crick worked, gave a talk at Guys Hospital Medical School in London on Thursday, May 14, 1953 which resulted in an article by Ritchie Calder in The News Chronicle of London, on Friday, May 15, 1953, entitled "Why You Are You.

Nearer Secret of Life." The news reached readers of Thc New York Times the next day; Victor K. McElheny, in researching his biography, "Watson and DNA: Making a Scientific Revolution", found a clipping of a six-paragraph New York Times article written from London and dated May 16, 1953 with the headline "Form of `Life Unit' in Cell Is Scanned."

The article ran in an early edition and was then pulled to make space for news deemed more important. (The New York Times subsequently ran a longer article on June 12, 1953). The Cambridge University undergraduate newspaper also ran its own short article on the discovery on Saturday, May 30, 1953.

Bragg's original announcement at a Solvay conference on proteins in Belgium on 8 April 1953 went unreported by the press! In 1962 Watson, Crick, and Maurice Wilkins jointly received the Nobel Prize for Physiology or Medicine for their determination of the structure of DNA.

"Central Dogma"

Watson and Crick's model attracted great interest immediately upon its presentation. Arriving at their conclusion on February 21, 1953, Watson and Crick made their first announcement on February 28. In an influential presentation in 1957, Crick laid out the "Central Dogma", which foretold the relationship between DNA, RNA, and proteins, and articulated the "sequence hypothesis."

A critical confirmation of the replication mechanism that was implied by the double-helical structure followed in 1958 in the form of the Meselson-Stahl experiment. Work by Crick and coworkers showed that the genetic code was based on non-overlapping triplets of bases, called codons, and Har Gobind Khorana and others deciphered the genetic code not long afterward (1966). These findings represent the birth of molecular biology.

History of RNA Tertiary Structure

Pre-history: The Helical Structure of RNA

The earliest work in RNA structural biology coincided, more or less, with the work being done on DNA in the early 1950s. In their seminal 1953 paper, Watson and Crick suggested that van der Waals crowding by the 2'OH group of ribose would preclude RNA from adopting a double helical structure identical to the model they proposed - what we now know as B-form DNA. This provoked questions about the three dimensional structure of RNA: could this molecule form some type of helical structure, and if so, how?

As with DNA, early structural work on RNA centered around isolation of native RNA polymers for fibre diffraction analysis. In part because of heterogeneity of the samples tested, early fibre diffraction patterns were usually ambiguous and not readily interpretable. In 1955, Grunberg-Manago *et al.* published a paper describing the enzyme polynucleotide phosphorylase, which cleaved a phosphate group from nucleotide diphosphates to catalyse their polymerization. This discovery allowed researchers to synthesize homogenous nucleotide polymers, which they then combined to produce double stranded molecules.

These samples yielded the most readily interpretable fibre diffraction patterns yet obtained, suggesting an ordered, helical structure for cognate, double stranded RNA that differed from that observed in DNA. These results paved the way for a series of investigations into the various properties and propensities of RNA. Through the late 1950s and early 1960s, numerous papers were publihed on various topics in RNA structure, including RNA-DNA hybridization, triple stranded RNA, and even small-scale crystallography of RNA di-nucleotides - G-C, and A-U - in primitive helix-like arrangements. For a more in-depth review of the early work in RNA structural biology.

The Beginning: Crystal Structure of tRNA

In the mid-1960s, the role of tRNA in protein synthesis was being intensively studied. At this point, ribosomes had been implicated in protein synthesis, and it had been shown that an mRNA strand was necessary for the formation of these structures.

In a 1964 publication, Warner and Rich showed that ribosomes active in protein synthesis contained tRNA molecules bound at the A and P sites, and discussed the notion that these molecules aided in the peptidyl transferase reaction. However, despite considerable biochemical characterization, the structural basis of tRNA function remained a mystery.

In 1965, Holley *et al.* purified and sequenced the first tRNA molecule, initially proposing that it adopted a cloverleaf structure, based largely on the ability of certain regions of the molecule to form stem loop structures. The isolation of tRNA proved to be the first major windfall in RNA structural biology. Following Holley's publication, numerous investigators began work on isolation tRNA for crystallographic study, developing improved methods for isolating the molecule as they worked.

By 1968 several groups had produced tRNA crystals, but these proved to be of limited quality and did not yield data at the resolutions necessary to determine structure. In 1971, Kim *et al.* achieved another breakthrough, producing crystals of yeast tRNA that diffracted to 2-3 Angstrom resolutions by using spermine, a naturally occurring polyamine, which bound to and stabilized the tRNA.

Despite having suitable crystals, however, the structure of tRNA was not immediately solved at high resolution; rather it took pioneering work in the use of heavy metal derivatives and a good deal more time to produce a high-quality density map of the entire molecule. In 1973, Kim *et al.* produced a 4 Angstrom map of the tRNA molecule in which they could unambiguously trace the entire backbone.

This solution would be followed by many more, as various investigators worked to refine the structure and thereby more thoroughly elucidate the details of base pairing and stacking interactions, and validate the published architecture of the molecule.

The tRNA structure is notable in the field of nucleic acid structure in general, as it represented the first solution of a long-chain nucleic acid structure of any kind - RNA or DNA - preceding Dickerson's solution of a B-form dodecamer by nearly a decade. Also, tRNA demonstrated many of the tertiary interactions observed in RNA architecture which would not be categorized and more thoroughly understood for years to come, providing a foundation for all future RNA structural research.

The Renaissance: the Hammerhead Ribozyme and the Group I Intron: $P_{4\text{-}6}$

For a considerable time following the first tRNA structures, the field of RNA structure did not dramatically advance. The ability to study an RNA strcuture depended upon the potential to isolate the RNA target. This proved limiting to the field for many years, in part owing to the fact that other know targets - i.e. the ribosome - were significantly more difficult to isolate and crystallize.

Further, because other interesting RNA targets had simply not been identified, or were not sufficiently understood to be deemed interesting, there was simply a lack of things to study structurally. As such, for some twenty years following the original publication of the tRNA structure, the structures of only a handful of other RNA targets were solved, with almost all of these belonging to the transfer RNA family. This unfortunate lack of scope would eventually be overcome largely because of two major advancements in nucleic acid research: the identification of ribozymes, and the ability to produce them via *in vitro* transcription.

Subsequent to Tom Cech's publication implicating the *Tetrahymena* group I intron as an autocatalytic ribozyme, and Sidney Altman's report of catalysis by ribonuclease P RNA, several other catalytic RNAs were identified in the late 1980s, including the hammerhead ribozyme. In 1994, McKay *et al.* published the structure of a 'hammerhead RNA-DNA ribozyme-inhibitor complex' at 2.6 Angstrom resolution, in which the autocatalytic activity of the ribozyme was disrupted via binding to a DNA substrate.

The conformation of the ribozyme published in this paper was eventually shown to be one of several possible states, and although this particular sample was catalytically inactive, subsequent structures have revealed its active-state architecture. This structure was followed by Doudna's publication of the structure of the P4-P6 domains of the *Tetrahymena* group I intron, a fragment of the ribozyme originally made famous by Cech.

The second clause in the title of this publication - *Principles of RNA Packing* - concisely evinces the value of these two structures: for the first time, comparisons could be made between well described tRNA structures and those of globular RNAs outside the transfer family. This allowed the framework of categorization to be built for RNA tertiary structure. It was now possible to propose the conservation of motifs, folds, and various local stabilizing interactions. For an early review of these structures and their implications.

In addition to the advances being made in global structure determination via crystallography, the early 1990s also saw the implementation of NMR as a powerful technique in RNA structural biology. Coincident with the large-scale ribozyme structures being solved crystallographically, a number of structures of small RNAs and RNAs complexed with drugs and peptides were solved using NMR. In addition, NMR was now being used to investigate and supplement crystal structures, as exemplified by the determination of an isolated tetraloop-receptor motif structure published in 1997. Investigations such as this enabled a more precise characterization of the base pairing and base stacking interactions which stabilized the global folds of large RNA molecules.

The importance of understanding RNA tertiary structural motifs was prophetically well described by Michel and Costa in their publication identifying the tetraloop motif: "..it should not come as a surprise if self-folding RNA molecules were to make intensive use of only a relatively small set of tertiary motifs. Identifying these motifs would greatly aid modelling enterprises, which will remain essential as long as the crystallization of large RNAs remains a difficult task".

The Modern Era: The Age of RNA Structural Biology

The resurgence of RNA structural biology in the mid-1990s has caused a veritable explosion in the field of nucleic acid structural research. Since the publication of the hammerhead and P_{4-6} structures, numerous major contributions to the field have been made. Some of the most noteworthy examples include the structures of the Group I and Group II introns, and the Ribosome. It should be noted that the first three structures were produced using *in vitro* transcription, and that NMR has played a role in investigating partial components of all four structures - testaments to the indispensability of both techniques for RNA research. Most recently, the 2009 Nobel Prize in Chemistry was awarded to Ada Yonath, Venkatraman Ramakrishnan, and Thomas Steitz for their structural work on the ribosome, demonstrating the prominent role RNA structural biology has taken in modern molecular biology.

History of Protein Biochemistry

First Isolation and Classification

Proteins were recognized as a distinct class of biological molecules in the eighteenth century by Antoine Fourcroy and others. Members of this class (called the "albuminoids", *Eiweisskorper*, or *matières albuminoides*) were recognized by their ability to coagulate or flocculate under various treatments such as heat or acid; well-known examples at the start of the nineteenth century included albumen from egg whites, blood serum albumin, fibrin, and wheat gluten. The similarity between the cooking of egg whites and the curdling of milk was recognized even in ancient times; for example, the name *albumen* for the egg-white protein was coined by Pliny the Elder from the Latin *albus ovi* (egg white).

With the advice of Jons Jakob Berzelius, the Dutch chemist Gerhardus Johannes Mulder carried out elemental analyses of common animal and plant proteins. To everyone's surprise, all proteins had nearly the same empirical formula, roughly $C_{400}H_{620}N_{100}O_{120}$ with individual sulfur and phosphorus atoms. Mulder published his findings in two papers (1837,1838) and hypothesized that there was one basic substance (*Grundstoff*) of proteins, and that it was synthesized by plants and absorbed from them by animals in digestion. Berzelius was an early proponent of this theory and proposed the name "protein" for this substance in a letter dated 10 July 1838

The name protein that I propose for the organic oxide of fibrin and albumin, I wanted to derive from [the Greek word], because it appears to be the primitive or principal substance of animal nutrition.

Mulder went on to identify the products of protein degradation such as the amino acid, leucine, for which he found a (nearly correct) molecular weight of 131 Da.

Purifications and Measurements of Mass

The minimum molecular weight suggested by Mulder's analyses was roughly 9 kDa, hundreds of times larger than other molecules being studied. Hence, the chemical structure of proteins (their primary structure) was an active area of research until 1949, when Fred Sanger sequenced insulin.

The (correct) theory that proteins were linear polymers of amino acids linked by peptide bonds was proposed independently and simultaneously by Franz Hofmeister and Emil Fischer at the same conference in 1902.

However, some scientists were sceptical that such long macromolecules could be stable in solution. Consequently, numerous alternative theories of the protein primary structure were proposed, e.g., the colloidal hypothesis that proteins were assemblies of small molecules, the cyclol hypothesis of Dorothy Wrinch, the diketopiperazine hypothesis of Emil Abderhalden and the pyrrol/ piperidine hypothesis of Troensgard (1942).

Most of these theories had difficulties in accounting for the fact that the digestion of proteins yielded peptides and amino acids. Proteins were finally shown to be macromolecules of well-defined composition (and not colloidal mixtures) by Theodor Svedberg using analytical ultracentrifugation.

The possibility that some proteins are non-covalent associations of such macromolecules was shown by Gilbert Smithson Adair (by measuring the osmotic pressure of hemoglobin) and, later, by Frederic M. Richards in his studies of ribonuclease S. The mass spectrometry of proteins has long been a useful technique for identifying posttranslational modifications and, more recently, for probing protein structure.

Most proteins are difficult to purify in more than milligram quantities, even using the most modern methods. Hence, early studies focused on proteins that could be purified in large quantities, e.g., those of blood, egg white, various toxins, and digestive/metabolic enzymes obtained from slaughterhouses. Many techniques of protein purification were developed during World War II in a project led by Edwin Joseph Cohn to purify blood proteins to help keep soldiers alive. In the late 1950s, the Armour Hot Dog Co. purified 1 kg (= one million milligrams) of pure bovine pancreatic ribonuclease A and made it freely available to scientists around the world. This generous act made RNase A the main protein for basic research for the next few decades, resulting in several Nobel Prizes.

Protein Folding and First Structural Models

The study of protein folding began in 1910 with a famous paper by Henrietta Chick and C. J. Martin, in which they showed that the flocculation of a protein was composed of two distinct processes: the precipitation of a protein from solution was *preceded* by another process called denaturation, in which the protein became much less soluble, lost its enzymatic activity and became more chemically reactive. In the mid-1920s, Tim Anson and Alfred Mirsky proposed that denaturation was a reversible process, a correct hypothesis that was initially lampooned by some scientists as "unboiling the egg". Anson also suggested that denaturation was a two-state ("all-or-none") process, in which one fundamental molecular transition resulted in the drastic changes in solubility, enzymatic activity and chemical reactivity; he further noted that the free energy changes upon denaturation were much smaller than those typically involved in chemical reactions.

In 1929, Hsien Wu hypothesized that denaturation was protein unfolding, a purely conformational change that resulted in the exposure of amino acid side chains to the solvent. According to this (correct) hypothesis, exposure of aliphatic and reactive side chains to solvent rendered the protein less soluble and more reactive, whereas the loss of a specific conformation caused the loss of enzymatic activity. Although considered plausible, Wu's hypothesis was not immediately accepted, since so little was known of protein structure and enzymology and other factors could account for the changes in solubility, enzymatic activity and chemical reactivity. In the early 1960s, Chris Anfinsen showed that the folding of ribonuclease A was fully reversible with no external cofactors needed, verifying the "thermodynamic hypothesis" of protein folding that the folded state represents the global minimum of free energy for the protein.

The hypothesis of protein folding was followed by research into the physical interactions that stabilize folded protein structures. The crucial role of hydrophobic interactions was hypothesized by Dorothy Wrinch and Irving

Langmuir, as a mechanism that might stabilize her cyclol structures. Although supported by J. D. Bernal and others, this (correct) hypothesis was rejected along with the cyclol hypothesis, which was disproven in the 1930s by Linus Pauling (among others).

Instead, Pauling championed the idea that protein structure was stabilized mainly by hydrogen bonds, an idea advanced initially by William Astbury (1933). Remarkably, Pauling's incorrect theory about H-bonds resulted in his *correct* models for the secondary structure elements of proteins, the alpha helix and the beta sheet. The hydrophobic interaction was restored to its correct prominence by a famous article in 1959 by Walter Kauzmann on denaturation, based partly on work by Kaj Linderstrom-Lang. The ionic nature of proteins was demonstrated by Bjerrum, Weber and Arne Tiselius, but Linderstrom-Lang showed that the charges were generally accessible to solvent and not bound to each other (1949).

The secondary and low-resolution tertiary structure of globular proteins was investigated initially by hydrodynamic methods, such as analytical ultracentrifugation and flow birefringence. Spectroscopic methods to probe protein structure (such as circular dichroism, fluorescence, near-ultraviolet and infrared absorbance) were developed in the 1950s. The first atomic-resolution structures of proteins were solved by X-ray crystallography in the 1960s and by NMR in the 1980s. As of 2006, the Protein Data Bank has nearly 40,000 atomic-resolution structures of proteins. In more recent times, cryo-electron microscopy of large macromolecular assemblies and computational protein structure prediction of small protein domains are two methods approaching atomic resolution.

Importance

An old standing dynamic problem is how DNA "self-replication" takes place in living cells that should involve transient uncoiling of supercoiled DNA fibres. Although DNA consists of relatively rigid, very large elongated biopolymer molecules called "fibres" or chains (that are made of repeating nucleotide units of four basic types, attached to deoxyribose and phosphate groups), its molecular structure *in vivo* undergoes dynamic configuration changes that involve dynamically attached water molecules and ions. Supercoiling, packing with histones in chromosome structures, and other such supramolecular aspects also involve *in vivo* DNA topology which is even more complex than DNA molecular geometry, thus turning molecular modelling of DNA into an especially challenging problem for both molecular biologists and biotechnologists.

Like other large molecules and biopolymers, DNA often exists in multiple stable geometries (that is, it exhibits conformational isomerism) and configurational, quantum states which are close to each other in energy on the potential energy surface of the DNA molecule.

Such varying molecular geometries can also be computed, at least in principle, by employing *ab initio* quantum chemistry methods that can attain high accuracy for small molecules, although claims that acceptable accuracy can be also achieved for polynuclelotides, as well as DNA conformations, were recently made on the basis of VCD spectral data. Such quantum geometries define an important class of *ab initio* molecular models of DNA whose exploration has barely started especially in connection with results obtained by VCD in solutions.

More detailed comparisons with such *ab initio* quantum computations are in principle obtainable through 2D-FT NMR spectroscopy and relaxation studies of polynucleotide solutions or specifically labelled DNA, as for example with deuterium labels.

In an interesting twist of roles, the DNA molecule itself was proposed to be utilized for quantum computing. Both DNA nanostructures as well as DNA 'computing' biochips have been built.

DNA Structure

The structure of DNA shows a variety of forms, both double-stranded and single-stranded. The mechanical properties of DNA, which are directly related to its structure, are a significant problem for cells. Every process which binds or reads DNA is able to use or modify the mechanical properties of DNA for purposes of recognition, packaging and modification. The extreme length (a chromosome may contain a 10 cm long DNA strand), relative rigidity and helical structure of DNA has led to the evolution of histones and of enzymes such as topoisomerases and helicases to manage a cell's DNA. The properties of DNA are closely related to its molecular structure and sequence, particularly the weakness of the hydrogen bonds and electronic interactions that hold strands of DNA together compared to the strength of the bonds within each strand.

Experimental techniques which can directly measure the mechanical properties of DNA are relatively new, and high-resolution visualization in solution is often difficult. Nevertheless, scientists have uncovered large amount of data on the mechanical properties of this polymer, and the implications of DNA's mechanical properties on cellular processes is a topic of active current research.

The DNA found in many cells can be macroscopic in length - a few centimetres long for each human chromosome. Consequently, cells must compact or "package" DNA to carry it within them. In eukaryotes this is carried by spool-like proteins known as histones, around which DNA winds. It is the further compaction of this DNA-protein complex which produces the well known mitotic eukaryotic chromosomes.

Recombinant DNA Technology

Gene Targeting (GT)

Gene targeting (GT) is a key technology for the rational, accurate and safe exploitation of plantsthrough genetic manipulation. Moreover, it offers the potential to completely knockout the expression of target genes or to make specific changes to gene function, objectives that cannot be achieved by conventional transgenesis. The ability to target DNA integration would permit the locus-specific integration of a transgene into a predetermined site of the host genome, avoiding the accidental inactivation of an endogenous gene localized at the insertion site or the unexpected expression profiles of the transgene itself, the so-called position effect. Systematic isolation and sequencing of genomic DNA flanking the insertion sites (known as FSTs or Flanking Sequence Tag offers the opportunity to rapidly characterize plants altered in a candidate gene sequence.

This approach is notably most useful in fully sequenced genomes such as in Arabidopsis thaliana. With 125 Mbp and 26,422 genes, the Arabidopsis genome shows very limited synteny with the 420–466 Mbp and 60,000 predicted genes of the rice genome. The recombination machinery has been well conserved throughout evolution, as an essential component of cell survival. In nature, homologous recombination is a DNA maintenance pathway that protects chromosomes against damage affecting both DNA strands, such as double strand breaks (DSBs) or interstrand crosslinks. DSB repair (DSBR) has been one of the most investigated homologous repair pathways see SHRIVASTAVA1, SHARMA2 AND KUMAR 2007 cited in Kumar and Sopory 2007).

Recently plastid genome transformation technique has gained prominence due to its better integration and less chances of random spread. The genome propagated by higher plant plastids, the plastome, is typically a double stranded DNA molecule of 130 to 160 kb. Over one hundred copies of this genome can be present in a single plastid. It is ideally represented as a circular monomer containing 2 inverted repeats, even though reality is more complex since linear and circular multimers have been frequently detected The complete sequence of this highly polyploidy genome is available for about 20 different species of angiosperms.

The first successful transformation of tobacco was performed using as marker a mutant plastid DNA fragment covering the 16SrRNA gene derived from a line resistant to spectinomycin and streptomycin. Major improvements in the selection process were soon obtained with the dominant aadA marker gene, inactivating spectinomycin or streptomycin. When fused to GFP, this marker can be used to track the selection process. Genes encoding resistance to kanamycin, nptII and more recently aphA-6 are also possible options, and could be more appropriate for some species.

There are three different fates for the external DNA to get integrated into the native genome. They are homologous recombination, illegitimate recombination or nonhomologous end joining, and single-strand annealing. Single-strand annealing (SSA), a third path of repair, requires the presence of repeated sequences on both sides of a break. After exonuclease degradation of the 5' ends, repair occurs by annealing of the two complementary sequences, a process leading to the loss of the genetic information contained between these repeats. With respect to the species preferential DSB repair pathway, HR but also IR mediates transgene integration. This second aspect explains the inefficiency of GT in higher plants, which use HR as a minor pathway of repair. Thus, despite the fact that transgene integration processes are still unclear in plants transgenic DNA would be preferentially integrated by end joining whether or not sharing homology within the host genome.

GM Crops

The U.S. Food and Drug Administration (FDA) has found no evidence to indicate that either ordinary plant deoxyribonucleic acid (DNA) or the DNA inserted into plants using bioengineering presents food safety problems. Nor are the small amounts of the newly expressed proteins likely to change dramatically the safety profile of the plant. If safety concerns should arise, however, they would most likely fall into one of three broad categories: allergens, toxins, or anti-nutrients. FDA has extensive experience in evaluating the safety of such substances in food.

It is important to note that the kinds of food safety testing typically conducted by developers of a bioengineered food crop to ensure that their foods meet all applicable requirements of the Food, Drug and Cosmetics Act (FD&C Act) address these potential concerns. In the event that something unexpected does occur, this testing provides a way to detect such changes at the developmental stage and defer marketing until any concern is resolved.

As Aforementioned, some of the Food Safety Concerns that could Arise Include

Allergens: Foods normally contain many thousands of different proteins. While the majority of proteins do not cause allergic reactions, virtually all known human allergens are proteins. Since genetic engineering can introduce a new protein into a food plant, it is possible that this technique could introduce a previously unknown allergen into the food supply or could introduce a known allergen into a "new" food.

Toxins: It is possible that a new protein, as introduced into a crop as a result of the genetic modification, could cause toxicity.

Anti-nutrients: It is possible that the introduction of anti-nutrients, such as molecules like phytic acid, could reduce essential dietary minerals such as phosphorus. The use of genetic engineering techniques could also result in unintended alterations in the amounts of substances normally found in a food, such as a reduction of Vitamin C or an increase in the concentration of a naturally occurring toxicant in the plant food.

Legal and Regulatory Issues

One important component in ensuring food safety is the U.S. regulatory structure. The FDA regulates bioengineered plant food in conjunction with the United States Department of Agriculture (USDA) and the Environmental Protection Agency (EPA). FDA has authority under the FD&C Act to ensure the safety of all domestic and imported foods for man or animals in the United States market. The exceptions to this are meat, poultry and certain egg products, which are regulated by USDA. The safety of animal drug residues in meat and poultry, however, is regulated by FDA. Pesticides, including those bioengineered into a food crop, are regulated primarily by EPA. USDA's Animal and Plant Health Inspection Service (APHIS) oversees the agricultural and environmental safety of planting and field testing bioengineered plants. Bioengineered foods and food ingredients must adhere to the same standards of safety under the FD&C Act that apply to their conventionally bred counterparts. This means that these products must be as safe as the traditional foods in the market. FDA has the power to remove a food from the market, or sanction those marketing the food if the food poses a risk to public health. It is important to note that the FD&C Act places a legal duty on developers to ensure that the foods they market to consumers are safe and comply with all legal requirements.

Area under the commercialization of genetically modified (GM), often called biotech crops continued to grow for the ninth consecutive year at a sustained double-digit growth rate of 20% in 2004 (James, 2004). The estimated global area of approved GM crops for 2005 was 90.0 million hectares with $4.70 billion global market value—based on the sale price of GM seed plus any technology fees that apply. The global value of the GM crop market is projected a more than $5.0 billion for 2005 (James, 2004). Commercialization of genetically modified (GM) crops continued to grow for the ninth consecutive year.

It reflects the substantial improvements in productivity, the environment, economics, health and social benefits realized by farmers, consumers and society. At the same time the growing controversy over GM food products increased interest in food labelling and identity preservation (IP) of GM crops. Hence, an IP system must be designed to provide assurances that the desired traits are present (or absent) in a product from the seed source, through all steps of production and delivery, to the end user. There are numerous regulatory issues related to GM crops. These include the testing and acceptance of new GM crops for commercial introduction, both domestically and internationally.

Nearly every country has different approaches and many have their own regulatory framework, with an intent to prevent cross-contamination of the conventional food and feed industries. IP tracking software is also available in market to ease the burdens associated with precise record-keeping requirements. The economics of IP has been calculated by various scientists depending on different applied IP systems. Niche-marketing opportunities will grow, because of the availability of GM crops and finally, IP of agricultural commodities from GM crops can provide greater choice and value desired by both agricultural producers and consumers.

Based on annual percentage growth in area, of the eight leading GM crop countries, India had the highest percentage year-on-year growth in 2004 with an increase of 400% in Bt cotton area over 2003, followed by Uruguay (200%), Australia (100%), Brazil (66%), China (32%), south Africa (25%), Canada (23%), Argentina (17%) and USA at 11%. India increased its area of approved Bt cotton, introduced only two years ago, from approximately 38,038 hectares in 2002/ 03 to 560,000 hectares in 2004/05 seasons with Bt coverage of 11.65% and approximately 300,000 small farmers benefited from Bt cotton.

Regulatory Issues Related to GM Crops

There are numerous regulatory issues related to GM crops. These include the testing and acceptance of new GM crops for commercial introduction and the introduction of food products containing ingredients from GM crops, both domestically and internationally. Nearly every country has different approaches and many have their own regulatory framework. Regulation is a very dynamic issue with changes being reviewed and proposed in many countries on an ongoing basis.

Numerous regulatory actions are consequently being proposed as governments react to consumer concerns and pressures. Several countries have or have proposed to create new agencies to specifically cover GM crops. Approaches range from cautious acceptance to attempts to ban (growing and even imports), at least for the foreseeable future, all crops and products with GM traits. Each is approaching the testing, introduction, and acceptance of GM crops in its own manner and on its own time schedule.

Abiotic and Biotic Resistance

Coat Protein Mediated Resistance

CP is an important structural protein as it not only protects the viral nucleic acid from degradation, but also plays an important role in virus infection. Its functions includes acquisition and transmission of virus by vectors, cell to cell and long distance spread of the virus in host plants, and for some viruses, it regulates one or more steps of virus replication.Coat Protein (CP) mediated resistance has been demonstrated for 17 groups of viruses, and so far this strategy has shown best promise. CP transgenes have been shown to be effective in preventing or reducing infection and diseases caused by homologous and closely related viruses (Gonsalves, et al., 1998). Coat protein-mediated protection has been reported for Tobacco mosaic virus, TMV,Tomato mosaic virus, ToMV, (Sa), Cucumber mosaic virus, CMV, Alfalfa mosaic virus, AlMV, Potato virus X, PVX, Potato virus Y, PVY, Potato leaf roll virus, PLRV, Papaya ringspot virus (PRSV) and a number of other viruses. CP-mediated resistance in Cantaloupe, Papaya, Potato, Squash and Tomato has been tested under the field conditions with fair degree of protection against most of the viruses.

Anti-HIV Agents among Desert Plants

Around 40 million people are affected due to the Human Immuno-deficiency Virus globally. During the past decades, a large number of anti-viral screening experiments on medicinal plant extracts have been reported and have led to the selection of several extracts active towards herpes viruses.

A promising result of a naturally occurring antiherpetic agent was given by n- docosanol (a natural 22 carbon saturated fatty alcohol) which is undergoing phase III clinical trials in patients. Clinical testing of the topical formulation, or systemic administration of drug suspensions has demonstrated a good therapeutic index, since high doses of n- docosanol do not elicit appreciable toxicity. The findings show that natural products are still potential sources in the search for new antiherpatic agents (Hattori et al., 1995). Various plant extracts used in Ayurvedic medicine for inhibitory effects on HIV virus have been studied (Hattori, personal communication). A large number of such plants occur in semi-arid and arid climate of Rajasthan. Acquired immunodeficiency syndrome (AIDS), the great pandemic of the second half of the 20th Century, is still a threatening disease world wide. Many research approaches are currently aimed at developing novel agents to arrest the replication of HIV through various targets.

These may include the inhibition of reverse transcriptase (RT), protease (PR), membrane fusion and integrase. HIV PR enzyme has been demonstrated to play an essential role in viral replication. A range of HIV PR inhibitors have been designed and applied in clinical trials such as Sanqunavir, Ritonavir and Indinavir. However, the development of drug resistance by virus, irrespective of the target, remains as an overwhelming problem in AIDS chemotherapy. Thus there is great need to search for and develop new and different anti-HIV candidates from plants and natural products are of considerable importance. In search for anti-HIV active agents from natural products, many attempts at screening traditional medicines have been made.

Biotic and Abiotic Stress

Environmental abiotic stress conditions, and especially drought and salinity, are currently the major factors which reduce crop yields worldwide leading to the fact that more than 800 million people are chronically undernourished.. The United Nations Environment Program estimates that approximately 20% of agricultural land and 50% of cropland in the world is saltstressed This salinity, in particular, is an increasing problem and nearly half of the area under irrigation, is at risk to be lost due o building up of salinity. Therefore genetic improvement of salt tolerance has become an urgent need for the future of agriculture in arid and semiarid regions.. One way of solving this problem would be breeding tolerant varieties of crop plants that can be grown on saline soils, but these breeding programs are time consuming and remained elusive. Hence, many metabolic changes are known to occur in plants subjected to salt stress, physiological parameters such as ionic relations have been suggested to be used as tolerance indicators since they can be related to salt tolerance mechanisms.

Biofuel Production

India has over 180 million of wasteland out of which 90 million ha is uncultivable. The degraded and denuded lands arise due to soil erosions as well as secondary salinizations. However Calotropis procera is a potential plant for bioenergy and biofuel production in semi arid regions of the country because it is able to grow on such lands. The plant has a growth potential of 2 dry tones to 40 dry tones per ha depending on the agro climatic conditions of it's growth. The plant has high level of regeneration potential and could be harvested up to 4 times a year. The plant yields valuable hydrocarbons which could be converted into diesel substitutes. The bio-diesel derived from Calotropis procera is free from NOx gases, S02 and Suspended Particulate Matter (SPM) and has high cetane value. Due to it's enormous potential for growth under adverse climatic conditions Calotropis procera is suggested as potential plant for bio-diesel production under semi-arid and arid conditions. Jatropha curcas also provides non-edible oil which could be converted into methyl ester with a gain of glycerine. The JME is used as mix in the ratio of 05, to 20 percent blend to fossil fuel diesel and in Germany the use of Rape seed methyl ester is increasing.

Soil Salinity

Soil salinity affects plant production in many parts of the world, particularly on irrigated land. NaCl is the predominant salt in most saline environments. Many crop species are sensitive to high concentrations of salt with negative impacts on agricultural production. Maize (Zea mays L.) is considered a moderately salt-sensitive plant.. Salt resistance of plants is a complex phenomenon that involves biochemical and physiological processes as well as morphological and developmental changes.. In addition to general osmotic stress, high concentrations of Na+ are toxic to maize and molecular mechanisms for salt resistance have not been fully identified or characterized (Zoerb et al 2004). The analysis of the plant's proteome is an important amendment to the analysis of the genome, because gene expression is altered under salinity stress.

The proteome, in contrast to the genome, is not static but rather dependent on a number of responses influenced by internal and external factors. The plant adaptation to environmental stress, such as soil salinity, is expected to have a strong influence on proteins. One approach to study the molecular mechanisms of plant responses to salinity is to use 2D polyacrylamide gel electrophoresis. Furthermore, the identification of differentially regulated proteins can lead to the identification of proteins and their corresponding genes which are involved in the physiology of salt resistance. The high resolution achieved by 2D gels and computer-assisted analysis of the differentially regulated proteins were used to examine those proteins whose synthesis was modulated by salt treatment and to quantify these changes. As far as we know, our investigations are the first to characterize the differentially expressed proteins from roots and shoots of maize after treatment with low and high salt stress. Plant material was an efficiently Na+-excluding maize inbred line developed in our laboratory. According to Munns the growth response to salt stress consists of two phases, first, a water-deficit that results from the relatively high solute concentrations in the root medium and, second, ion-specific stress resulting from altered K+/Na+ ratios or Na+ and Cl" concentrations that are toxic to plants. The aim of this study was to elucidate biochemical and physiological reactions of glycophytes to salt stress in the first phase of salinity.

While all major crops, as well as most wild species, are glycophytes, i.e. sensitive to relatively low salt concentrations, there are plants naturally adapted to conditions of high salinity in the soil. These plants, known as halophytes, include a large taxonomic variety and occupy diverse habitats, from extreme dry to temporarily waterlogged sites or salt marshes, and can tolerate NaCl concentrations similar, or even higher than that of sea water, ca. 500 mM. It is - without doubt - necessary to develop sustainable biological production systems which can tolerate higher water salinity because freshwater resources will not come up with increasing demands of agricultural practice in near future. The sustainable use of halophytic plants is a promising approach to valorize strongly salinised zones unsuitable for conventional agriculture and mediocre waters. The development of cash crop halophytes and the breeding of salt resistant crop varieties will require a clear understanding of the complex mechanisms of salt stress tolerance, which we are still lacking despite intensive research during the last decade.

Recombinant Proteins

It has been estimated that 1 g of recombinant antibody could be produced in leaves of a plant crop for only about US$100 while the current prices for monoclonal antibodies range from US$2000 to US$5000 per gram. Indeed the cost of producing 1 kg of recombinant protein from most major field crops is estimated to be 10 to 50 times lower than the cost of producing the same amount by E. coli fermentation. Whole plants also have an advantage when tissues such as a fruit, tuber, etc. can be used to express the protein of interest (James and Lee 2001), and an area of undisputed advantage occurs when the oral delivery of pharmaceuticals, as well as feed and food enzymes, is possible. However, there are also some evident obstacles that arise when a whole plant is used for large-scale protein production.

Cell-free systems have proved to have high utility at the genomic, transcriptomic and proteomic levels and to form a vital component of many aspects of recombinant gene expression, and of both structural and functional proteomics..

Compared with DNA microarrays, protein bio-chips provide more challenges and have yet to be perfected due to the complexity and inherent difficulties with protein immobilization.

Novel Cell Free Translation System is Unique Discovery

A novel cell-free translation system is described in which template-mRNA molecules were captured onto solid surfaces to simultaneously synthesize and immobilize proteins in a more native-state form. This technology comprises a novel solid-phase approach to cell-free translation and RNA–protein fusion techniques.

A newly constructed biotinylated linker-DNA which enables puromycin-assisted RNA–protein fusion is ligated to the 32 ends of the mRNA molecules to attach the mRNA-template on a streptavidin-coated surface and further to enable the subsequent reactions of translation and RNA–protein fusion on surface. The protein products are therefore directly immobilized onto solid surfaces and furthermore were discovered to adopt a more native state with proper protein folding and superior biological activity compared with conventional liquid-phase approaches. We further validate this approach via the production of immobilized green fluorescent protein (GFP) on microbeads and by the production and assay of aldehyde reductase (ALR) enzyme with 4-fold or more activity. The approach developed in this study may enable to embrace the concept of the transformation of 'RNA chip-to-protein chip' using a solid-phase cell-free translation system and thus to the development of high-throughput microarray platform in the field of functional genomics and in vitro evolution.

Plant Tissue Culture

Another area of biotechnology is micropropagation of plants. The aim of this technique is a fast production of a great number of genetically identical plants from a highly valuable mother plant or e.g. monosexual male and female plants. These plants can be either directly sold on the market for planting, used for breeding purposes, for genetechnology or the technique is used as a method for basic science studies. Using petiole explants from transgenic plants containing the auxin responsive MAS promoter linked to the GUS reporter gene the distribution of auxin within the cultured petiole could be followed during the induction phase of somatic embryogenesis. Interestingly, the cells forming the glandular canal contain high concentrations of auxins as shown by using transgenic plants containing the auxin sensitive MAS-promoter coupled to the GUS-gene, whatever the significance. Rhizogenic centres develop near vascular bundles prior to those embryogenic centres.

Genetic factors play a central role to induce somatic embryos, i.e. to provide the competence of the species for the process. Here, great variation can be found even within a genera such as Daucus. Eight of twelve Daucus species cultured in identical conditions produced somatic embryos (D. halophilus, D. capillifolius, D. commutatus, D. azoricus, D. gadacei, D. maritimus, D. maximus, D. carota), whereas four species (D. montevidensis, D. pussillus, D. muricatus, D. glochidiatus) were not competent to do so. Under identical culture conditions, only 8 out of 12 species and subspecies of the genus Daucus proved capable of somatic embryogenesis. Random amplified polymorphic DNA analysis indicated a polymorphism between the genomes of individual species that were capable of embryogenesis and those that were not. Two specific bands (1.1 kbp, 0.68 kbp) were detected only inthe genomes of individuals with the capacity for embryogenesis. These were cloned and sequenced, and the homology of the nucleotide sequences of the various species was detected: this ranged from 74% to 92% for the larger sequence and from 92% to 97% for the smaller one. These DNA sequences would appear to be useful as a marker of the capacity for somatic embryogenesis in the genus Daucus (Imani et al.2001) The sequences obtained in this study have been registered with the European Bioinformatics Institute (EMBL). The access numbers for the sequences are: AJ278039 DCA78039; AJ278040 DCA78040; AJ278041 CA78041; AJ278042 DCA78042; AJ278176 DCA278176; AJ278177 DCA278177; AJ278178 DCA278178; AJ278179 DCA278179. No open reading frames were detected.

We performed later additional studies with other Daucus species (D. capillifolius; D. carota ssp. Azoricus and gadecaei) to determine the use-fulness of these RAPD products as markers for identifying the ability of Daucus species to generate somatic embryos. There was a 100% correlation between the embryogenic potential of the species and the occurrence of the 1.1-kbp and

0.68-kbp band (Imani et al.2001). Micropropagation Technique in Enhancing the Productivity of Crops have been taken up at large scale at TERI. Some of the activities undertaken at MTP include:

- Large-scale production of superior quality planting material of various economically important plant species using tissue culture technology
- Mass multiplication of those species which are difficult to regenerate by conventional methods of propagation or where conventional methods of propagation are inadequate to meet the demand of planting material
- Development of new micropropagation protocols and refining of others so as to make them suitable for large-scale production of plants
- Helping the entrepreneurs/industry through technology transfer, mother cultures and training of staff
- Assisting clients in setting-up their own tissue culture labs
- Creating awareness.

Till date over 15 million plants of forest species, cash crops, medicinal and aromatic plants, and ornamentals (foliage and flowering) have been dispatched to various state forest and horticulture departments, private entrepreneurs, nurseries, farmers etc. for field demonstrations and routine plantations. In addition, MTP is in possession of micropropagation protocols for over 90 economically important plant species. Field demonstration plots of tissue cultured plants have been laid at different locations to evaluate and compare their growth performances with conventional plants. Besides transfer of technologies to industry for commercialization, MTP has been instrumental in capacity building and creating awareness about tissue culture technology through seminars/ workshops/training programmes, exhibitions, etc.

2

DNA Nanotechnology

DNA nanotechnology is a branch of nanotechnology which uses the molecular recognition properties of DNA and other nucleic acids to create designed, artificial structures out of DNA for technological purposes. In this field, DNA is used as a structural material rather than as a carrier of genetic information, making it an example of bionanotechnology. DNA nanotechnology has applications in molecular self-assembly and in DNA computing.

Although DNA is usually considered in the context of molecular biology, as the carrier of genetic information in living cells, DNA nanotechnology considers DNA solely as a chemical and as a material, and is usually pursued outside of any biological context. DNA nanotechnology makes use of the fact that, due to the specificity of Watson-Crick base pairing, only portions of the strands which are complementary to each other will bind to each other to form duplex DNA. DNA nanotechnology attempts to rationally design sets of DNA strands so that desired portions of each strand will assemble in the correct positions to for some desired target structure.

Although the field is usually called DNA nanotechnology, its principles apply equally well to other nucleic acids such as RNA and PNA, and structures incorporating these have been made. For this reason the field is occasionally referred to as nucleic acid nanotechnology.

Fundamental Concepts

DNA nanotechnology creates complex structures out of nucelic acids by making use of the specificity of base pairing in nucelic acid molecules. The structure of a nucleic acid molecule consists of a sequence of nucleotides, distinguished by which nucleobase they contain. In DNA, the four bases used are adenine (A), cytosine (C), guanine (G), and thymine (T). Nucleic acids have the property that two molecules will bind to each other to form a double helix only if the two sequences are complementary, meaning that they form matching sequences of base pairs, with A's only binding to T's, and C's only to G's.

Beacuse the formation of correctly matched base pairs is energetically favourable, nucleic acid strands are expected in most cases to bind to each other in the conformation that maximizes the number of correctly paired bases. This property, that the sequence determines the pattern of binding and the overall structure, is used by the field of DNA nanotechnology in that sequences are rationally designed so that a desired structure is favoured to form.

Nearly all structures in DNA nanotechnology make use of branched DNA structures containing junctions, as opposed to most biological DNA which exists in a linear double helix form. One of the simplest branched structures, and the first made, is a four-arm junction which can be made using four individual DNA strands which are complementary to each other in the correct pattern. Unlike in natural Holliday junctions, in the artificial immobile four-arm junction shown below, the base sequence of each arm is different, meaning that the junction point is fixed in a certain position.

Junctions can be used in more complex molecules. One of the more widely-used of these is the "double-crossover" or DX motif. A DX molecule can be thought of as two DNA duplexes positioned parallel to each other, with two crossover points where strands cross from one duplex into the other. Each junction point is itself topologically a four-arm junction.

This molecule has the advantage that the junction points are now constrained to a single orientation as opposed to being flexible as in the four-arm junction. This makes the DX motif suitable as a structural building block for larger DNA complexes.

Design

DNA nanostructures must be designed so that they will assemble into the desired structures. This includes both the design of secondary structure, deciding which parts of which nucleic acid molecules should bind to each other, and primary structure, the specification of the identity of each individual base.

Structural Design

The first step in a designing a nucleic acid nanostructure is to decide how a given structure should be represented by a specific arrangement of nucleic acid strands. This design step thus determines the secondary structure of the nucleic acid complex which will assemble into the desired shape. There are several approaches which have been demonstrated:

- Sequence symmetry minimization. Most design in DNA nanotechnology focuses on designing sequences so that the target structure is a thermodynamic minimum, and mis-assembled structures have higher energies and are thus disfavored.
- Folding structures. An alternative to the tile-based approach, two-dimensional DNA structures can be made from a single, long DNA strand of arbitrary sequence which is folded into the desired shape by using shorter, "staple" strands. This allows the creation of two-dimensional shapes at the nanoscale using DNA. Demonstrated designs have included the smiley face and a coarse map of North America. DNA origami was the cover story of *Nature* on March 15, 2006.
- Kinetic assembly. Recently, there has been interest in controlling the kinetics of DNA self-assembly, so that transient dynamics can also be programmed into the assembly. Such a method also has the advantage of proceeding isothermally and thus not requiring a thermal annealing step required by solely thermodynamic approaches.

Nucleic Acid Design

Nucleic acid design is the process of generating a set of nucleic acid base sequences that will associate into a desired conformation. Nucleic acid design is central to the fields of DNA nanotechnology and DNA computing.

It is necessary because there are many possible sequences of nucleic acid strands that will fold into a given secondary structure, but many of these sequences will have undesired additional interactions which must be avoided. In addition, there are many tertiary structure considerations which affect the choice of a secondary structure for a given design.

Nucleic acid design has similar goals to protein design: in both, the sequence of monomers is designed to favor the desired folded or associated structure and to disfavour alternate structures.

However, nucleic acid design has the advantage of being a much computationally simpler problem, since the simplicity of Watson-Crick base pairing rules leads to simple heuristic methods which yield experimentally robust designs. Computational models for protein folding require tertiary structure information whereas nucleic acid design can operate largely on the level of secondary structure. However, nucleic acid structures are less versatile than proteins in their functionality.

Nucleic acid design can be considered the inverse of nucleic acid structure prediction. In structure prediction, the structure is determined from a known sequence, while in nucleic acid design, a sequence is generated which will form a desired structure.

Fundamental Concepts

The structure of nucleic acids consists of a sequence of nucleotides. There are four types of nucleotides distinguished by which of the four nucleobases they contain: in DNA these are adenine (A), cytosine (C), guanine (G), and thymine (T). Nucleic acids have the property that two molecules will bind to each other to form a double helix only if the two sequences are complementary, that is, they can form matching sequences of base pairs. Thus, in nucleic acids the sequence determines the pattern of binding and thus the overall structure. Nucleic acid design is the process by which, given a desired target structure or functionality, sequences are generated for nucleic acid strands which will self-assemble into that target structure. Nucleic acid design encompasses all levels of nucleic acid structure:

- Primary structure—the raw sequence of nucleobases of each of the component nucleic acid strands;
- Secondary structure—the set of interactions between bases, i.e., which parts of which strands are bound to each other; and
- Tertiary structure—the locations of the atoms in three-dimensional space, taking into consideration geometrical and steric constraints.

One of the greatest concerns in nucleic acid design is ensuring that the target structure has the lowest energy (i.e. is the most thermodynamically favourable) whereas misformed structures have higher energy and are thus unfavored. These goals can be achieved through the use of a number of apporaches, including heuristic, thermodynamic, and geometrical ones. Almost all nucleic acid design tasks are aided by computers, and a number of software packages are available for many of these tasks.

Two considerations in nucleic acid design are that desired hybridizations should have melting temperatures in a narrow range, and any spurious interactions should have very low melting temperatures (i.e. they should be very weak). There is also a contrast between affinity-optimizing "positive design", seeks to minimize the energy of the desired structure in an absolute sense, and specificity-optimizing "negative design", which considers the energy of the target structure relative to those of undesired structures. Algorithms which implement both kinds of design tend to perform better than those that consider only one type.

Approaches

Heuristic Methods

Heuristic methods use simple criteria which can be quickly evaluated to judge the suitability of different sequences for a given secondary structure. They have the advantage of being much less computationally expensive than the energy minimization algorithms needed for thermodynamic or geometrical modelling, and being easier to implement, but at the cost of being less rigorous than these models.

Sequence symmetry minimization is the oldest approach to nucleic acid design and was first used to design immobile versions of branched DNA structures. Sequence symmetry minimization divides the nucleic acid sequence into overlapping subsequences of a fixed length, called the criterion length. Each of the 4^N possible subsequences of length N is allowed to appear only once in the sequence. This ensures that no undesired hybridizations can occur which have a length greater than or equal to the criterion length.

A related heuristic approach is to consider the "mismatch distance", meaning the number of positions in a certain frame where the bases are not complementary. A greater mismatch distance lessens the chance that a strong spurious interaction can happen. This is related to the concept of Hamming distance in information theory. Another related but more involved approach is to use methods from coding theory to construct nucleic acid sequences with desired properties.

Thermodynamic Models

Information about the secondary structure of a nucleic acid complex along with its sequence can be used to predict the thermodyamic properties of the complex.

When thermodynamic models are used in nucleic acid design, there are usually two considerations: desired hybridizations should have melting temperatures in a narrow range, and any spurious interactions should have very low melting temperatures (i.e. they should be very weak). The free energy of a perfectly matched nucleic acid duplex can be predicted using a nearest neighbour model.

This model considers only the interactions between a nucleotide and its nearest neighbours on the nucleic acid strand, by summing the free energy of each of the overlapping two-nucleotide subwords of the duplex. This is then corrected for self-complementary monomers and for GC-content. Once the free energy is known, the melting temperature of the duplex can be determined. GC-content alone can also be used to estimate the free energy and melting temperature of a nucleic acid duplex. This is less accurate but also much less computationally costly.

Software for thermodynamic modelling of nucleic acids includes Nupack, mfold, and Vienna. A related approach, inverse secondary structure prediction, uses stochastic local search which improves a nucleic acid sequence by running a structure prediction algorithm and the modifying the sequence to eliminate unwanted features.

Geometrical Models

Geometrical models of nucleic acids are used to predict tertiary structure. This is important because designed nucleic acid complexes usually contain multiple junction points, which introduces geometric constraints to the system. These constraints stem from the basic structure of nucleic acids, mainly that the double helix formed by nucleic acid duplexes has a fixed helicity of about 10.4 base pairs per turn, and is relatively stiff.

Because of these constraints, the nucleic acid complexes are sensitive to the relative orientation of the major and minor grooves at junction points. Geometrical modelling can detect strain stemming from misalignments in the structure, which can then be corrected by the designer.

Geometric models of nucleic acids for DNA nanotechnology generally use reduced representations of the nucleic acid, because simulating every atom would be very computationally expensive for such large systems. Models with three pseudo-atoms per base pair, representing the two backbone sugars and the helix axis, have been reported to have a sufficient level of detail to predict experimental results. However, models with five pseudo-atoms per base pair, explicitly including the backbone phosphates, are also used.

Software for geometrical modelling of nucleic acids includes GIDEON, Tiamat, Nanoengineer-1, and UNIQUIMER 3D. Geometrical concerns are especially of interest in the design of DNA origami, because the sequence is predetermined by the choice of scaffold strand. Software specifically for DNA origami design has been made, including caDNAno and SARSE.

Applications

Nucleic acid design is used in DNA nanotechnology to design strands which will self-assemble into a desired target structure. These include examples such as DNA machines, periodic two- and three-dimensional lattices, polyhedra, and DNA origami. It can also be used to create sets of nucleic acid strands which are "orthogonal", or non-interacting with each other, so as to minimize or eliminate spurious interactions. This is useful in DNA computing, as well as for molecular barcoding applications in chemical biology and biotechnology.

Types of Structures

Many structures made from DNA have been synthesized and characterized.

Periodic Lattices

The earliest method for creating DNA nanostructures was to construct them out of smaller discrete units. This method has the advantage of being able to conceptually separate the stronger interactions which form each tile from the assembly of the larger complete structure. It is often used to make periodic lattices, but can also be used to implement algorithmic self-assembly, making them one platform for DNA computing. DX, or Double Crossover, molecules can be equipped with sticky ends in order to combine them into a two-dimensional periodic lattice. Each DX molecule has four termini, one at each end of the two double-helical domains, and these can be equipped with sticky ends that program them to combine into a specific pattern.

More than one type of DX can be used which can be made to arrange in rows or any other tessellated pattern. They thus form extended flat sheets which are essentially two-dimensional crystals of DNA. Two-dimensional arrays have been made out of other motifs as well, including the Holliday junction rhombus array as well as various DX-based arrays in the shapes of triangles and hexagons.

Creating three-dimensional lattices out of DNA was the earliest goal of DNA nanotechnology, but proved to be one of the most difficult to realize. Success in constructing three-dimensional DNA lattices was finally reported in 2009 using a motif based on the concept of tensegrity, a balance between tension and compression forces.

Nanotubes

In addition to flat sheets, DX arrays have been made to form hollow nanotubes of 4-20 nm diameter. These DNA nanotubes are somewhat similar in size and shape to carbon nanotubes, but the carbon nanotubes are stronger and better conductors, whereas the DNA nanotubes are more easily modified and connected to other structures.

Polyhedra

A number of three-dimensional DNA molecules have been made which have the connectivity of a polyhedron such as an octahedron or cube. In other words, the DNA duplexes trace the edges of a polyhedron with a DNA junction at each vertex.

The earliest demonstrations of DNA polyhedra involved multiple ligations and solid-phase synthesis steps to create catenated polyhedra. More recent work has yielded polyhedra whose synthesis is much easier. These include a DNA octahedron made from a long single strand designed to fold into the correct conformation, as well as a tetrahedron which can be produced from four DNA strands in a single step.

Arbitary Shapes

DNA structures with solid faces have also been constructed, using the DNA origami method. These can be programmed to open and release their cargo in response to a stimulus, making them potentially useful as programmable molecular cages.

Functional Nucleic Acid Nanostructures

DNA nanotechnology focuses on creating molecules with designed functionalities as well as structures. Many classes of functional systems have been demonstrated.

Nanoarchitecture

The idea of using DNA arrays to template the assembly of other functional molecules was first suggested by Nadrian Seeman in 1987, but only recently has progress been made in reducing these kinds of schemes to practice. In 2006, researchers covalently attached gold nanoparticles to a DX-based tile and showed that self-assembly of the DNA structures also assembled the nanoparticles hosted on them. A non-covalent hosting scheme was shown in 2007, using Dervan polyamides on a DX array to arrange streptavidin proteins on specific kinds of tiles on the DNA array.

Previously in 2006, Dwyer and LaBean demonstrated the letters "D" "N" and "A" created on a 4x4 DX array using streptavidin. In 2007, a hierarchical assembly based on this approach was also demonstrated that scales to larger arrays (8X8 and 8.96 MD).

There has also been interest in using DNA nanotechnology to assemble molecular electronics devices. To this end, DNA has been used to assemble single walled carbon nanotubes into field-effect transistors.

DNA Computing

DNA computing is a form of computing which uses DNA, biochemistry and molecular biology, instead of the traditional silicon-based computer technologies. DNA computing, or, more generally, biomolecular computing, is a fast developing interdisciplinary area. Research and development in this area concerns theory, experiments and applications of DNA computing.

History

This field was initially developed by Leonard Adleman of the University of Southern California, in 1994. Adleman demonstrated a proof-of-concept use of DNA as a form of computation which solved the seven-point Hamiltonian path problem. Since the initial Adleman experiments, advances have been made and various Turing machines have been proven to be constructible.

In 2002, researchers from the Weizmann Institute of Science in Rehovot, Israel, unveiled a programmable molecular computing machine composed of enzymes and DNA molecules instead of silicon microchips. On April 28, 2004, Ehud Shapiro, Yaakov Benenson, Binyamin Gil, Uri Ben-Dor, and Rivka Adar at the Weizmann Institute announced in the journal Nature that they had constructed a DNA computer coupled with an input and output module which would theoretically be capable of diagnosing cancerous activity within a cell, and releasing an anti-cancer drug upon diagnosis. In 2009, biocomputing systems were coupled with standard silicon based chips for the first time.

In this experiment, an enzyme based OR-Reset/AND-Reset logic system was achieved using field-effect Silicon chips. This advancement could yield great potential in the fields of Synthetic Biology, and Biomedical Engineering, as it marks the integration of biological and electro-mechanical systems on a sub-cellular level.

Capabilities

DNA computing is fundamentally similar to parallel computing in that it takes advantage of the many different molecules of DNA to try many different possibilities at once.

DNA computing also offers much lower power consumption than traditional silicon computers. DNA uses adenosine triphosphate (ATP) as fuel to allow ligation or as a means to heat the strand to cause disassociation. Both strand hybridization and the hydrolysis of the DNA backbone can occur spontaneously, powered by the potential energy stored in DNA.

Consumption of two ATP molecules releases 1.5 x 10^{e10} J. Even with a large number of transitions per second using two ATP molecules, power output is still low. For instance, Kahan reports 109 transitions per second with an energy consumption of 10^{e10} W, and similarly Shapiro reports a system producing 7.5 x 10^{11} outputs in 4000 sec resulting in an energy consumption rate of ~ 10^{e10} W.

For certain specialized problems, DNA computers are faster and smaller than any other computer built so far. Furthermore, particular mathematical computations have been demonstrated to work on a DNA computer. As an example, Aran Nayebi has provided a general scalable implementation of Strassen's matrix multiplication algorithm on a DNA computer.

But DNA computing does not provide any new capabilities from the standpoint of computability theory, the study of which problems are computationally solvable using different models of computation.

For example, if the space required for the solution of a problem grows exponentially with the size of the problem (EXPSPACE problems) on von Neumann machines, it still grows exponentially with the size of the problem on DNA machines. For very large EXPSPACE problems, the amount of DNA required is too large to be practical. (Quantum computing, on the other hand, *does* provide some interesting new capabilities.) DNA computing overlaps with, but is distinct from, DNA nanotechnology. The latter uses the specificity of Watson-Crick basepairing and other DNA properties to make novel structures out of DNA.

These structures can be used for DNA computing, but they do not have to be. Additionally, DNA computing can be done without using the types of molecules made possible by DNA nanotechnology.

Methods

There are multiple methods for building a computing device based on DNA, each with its own advantages and disadvantages. Most of these build the basic logic gates (AND, OR, NOT) associated with digital logic from a DNA basis. Some of the different bases include DNAzymes, deoxyoligonucleotides, enzymes, DNA tiling, and polymerase chain reaction.

DNAzymes

Catalytic DNA (deoxyribozyme or DNAzyme) catalyse a reaction when interacting with the appropriate input, such as a matching oligonucleotide. These DNAzymes are used to build logic gates analogous to digital logic in silicon; however, DNAzymes are limited to 1-, 2-, and 3-input gates with no current implementation for evaluating statements in series.

The DNAzyme logic gate changes its structure when it binds to a matching oligonucleotide and the fluorogenic substrate it is bonded to is cleaved free. While other materials can be used, most models use a fluorescence-based substrate because it is very easy to detect, even at the single molecule limit. The amount of fluorescence can then be measured to tell whether or not a reaction took place.

The DNAzyme that changes is then "used," and cannot initiate any more reactions. Because of this, these reactions take place in a device such as a continuous stirred-tank reactor, where old product is removed and new molecules added. Two commonly used DNAzymes are named E6 and 8-17. These are popular because they allow cleaving of a substrate in any arbitrary location. Stojanovic and MacDonald have used the E6 DNAzymes to build the MAYA I and MAYA II machines, respectively; Stojanovic has also demonstrated logic gates using the 8-17 DNAzyme. While these DNAzymes have been demonstrated to be useful for constructing logic gates, they are limited by the need for a metal cofactor to function, such as Zn^{2+} or Mn^{2+}, and thus are not useful in vivo.

A design called a *stem loop*, consisting of a single strand of DNA which has a loop at an end, are a dynamic structure that opens and closes when a piece of DNA bonds to the loop part. This effect has been exploited to create several logic gates. These logic gates have been used to create the computers MAYA I and MAYA II which can play tic-tac-toe to some extent.

Enzymes

Enzyme based DNA computers are usually of the form of a simple Turing machine; there is analogous hardware, in the form of an enzyme, and software, in the form of DNA. Shapiro has demonstrated a DNA computer using the FokI enzyme and expanded on his work by going on to show automata that diagnose and react to prostate cancer: under expression of the genes PPAP2B and GSTP1 and an over expression of PIM1 and HPN.

His automata evaluated the expression of each gene, one gene at a time, and on positive diagnosis then released a single strand DNA molecule (ssDNA) that is an antisense for MDM2. MDM2 is a repressor of protein 53, which itself is a tumour suppressor. On negative diagnosis it was decided to release a suppressor of the positive diagnosis drug instead of doing nothing. A limitation of this implementation is that two separate automata are required, one to administer each drug.

The entire process of evaluation until drug release took around an hour to complete. This method also requires transition molecules as well as the FokI enzyme to be present. The requirement for the FokI enzyme limits application *in vivo*, at least for use in "cells of higher organisms". It should also be pointed out that the 'software' molecules can be reused in this case.

Toehold Exchange

DNA computers have also been constructed using the concept of toehold exchange. In this system, an input DNA strand binds to a sticky end, or toehold, on another DNA molecule, which allows it to displace another strand segment from the molecule. This allows the creation of modular logic components such as AND, OR, and NOT gates and signal amplifiers, which can be linked into arbitrarily large computers. This class of DNA computers does not require enzymes or any chemical capability of the DNA.

Algorithmic Self-assembly

DNA nanotechnology has been applied to the related field of DNA computing. DNA tiles can be designed to contain multiple sticky ends with sequences chosen so that they act as Wang tiles. A DX array has been demonstrated whose assembly encodes an XOR operation; this allows the DNA array to implement a cellular automaton which generates a fractal called the Sierpinski gasket. This shows that computation can be incorporated into the assembly of DNA arrays, increasing its scope beyond simple periodic arrays.

DNA Machine

A DNA machine is a molecular machine constructed from DNA. Research into DNA machines was pioneered in the late 1980s by Nadrian Seeman and co-workers from New York University. DNA is used because of the numerous biological tools already found in nature that can affect DNA, and the immense knowledge of how DNA works previously researched by biochemists.

DNA machines can be logically designed since DNA assembly of the double helix is based on strict rules of base pairing that allow portions of the strand to be predictably connected based on their sequence. This 'selective stickiness' is a key advantage in the construction of DNA machines.

An example of a DNA machine was reported by Bernard Yurke and co-workers at Lucent Technologies in the year 2000, who constructed molecular tweezers out of DNA. The DNA tweezers contain three strands: A, B and C. Strand A latches onto half of strand B and half of strand C, and so it joins them all together. Strand A acts as a hinge so that the two 'arms' — AB and AC — can move. The structure floats with its arms open wide. They can be pulled shut by adding a fourth strand of DNA (D) 'programmed' to stick to both of the dangling, unpaired sections of strands B and C. The closing of the tweezers was proven by tagging strand A at either end with light-emitting molecules that do not emit light when they are close together. To re-open the tweezers add a further strand (E) with the right sequence to pair up with strand D. Once paired up, they have no connection to the machine BAC, so float away. The DNA machine can be opened and closed repeatedly by cycling between strands D and E. These tweezers can be used for removing drugs from inside fullerenes as well as from a self assembled DNA tetrahedron. The state of the device can be determined by measuring the separation between donor and acceptor fluorophores using FRET.

Materials and Methods

DNA of custom sequence is readily available through oligonucleotide synthesis. This process is usually automated by using a DNA synthesizing machine, and custom DNA is commercially available from many vendors. The sequences of the individual DNA strands which make up the target structure are designed computationally. Molecular modelling and thermodynamic modelling are sometimes used to optimize the DNA sequences. The DNA molecules created by DNA nanotechnology are usually characterized by gel electrophoresis, which provides information about the size and shape of DNA molecules, indicating whether they have formed properly. Fluorescent labelling and Fluorescence resonance energy transfer (FRET) are also used to characterize the structure of the molecules.

DNA structures can be directly imaged by atomic force microscopy, which images structures deposited on a flat surface. This method is well-suited to extended two-dimensional structures, but is less useful for discrete three-dimensional structures. For these latter structures transmission electron microscopy and cryo-electron microscopy are important methods. Extended three-dimensional lattices are analysed by X-ray crystallography. Kinetics of DNA self assembly can be studied by real time techniques such as dual polarisation interferometry and QCMD.

History

The concept of DNA nanotechnology was invented by Nadrian Seeman in the early 1980s. Seeman was originally concerned with using a three-dimensional DNA lattice to orient target molecules, which would simplify their crystallographic study by eliminating the difficult process of obtaining pure crystals.

This idea had reportedly come to him in fall 1980, after realizing the similarity between the M. C. Escher woodcut *Depth* and an array of DNA six-arm junctions. To this end, Seeman's laboratory in 1991 published the synthesis of a cube made of DNA, the first three-dimensional nanoscale object, for which he received the 1995 Feynman Prize in Nanotechnology, which was followed by a DNA truncated octahedron. However, it soon became clear that these molecules, polygonal shapes with flexible junctions as their vertices, were not rigid enough to form extended three-dimensional lattices.

Seeman developed the more rigid double-crossover (DX) motif, and in collaboration with Erik Winfree, in 1998 published the creation of two-dimensional lattices of DX tiles. These tile-based structures had the advantage that they provided the capability to implement DNA computing, which was demonstrated by Winfree and Paul Rothemund in 2004, and for which they shared the 2006 Feynman Prize in Nanotechnology.

The field has continued to branch out. The first DNA nanomachine—a motif which changes its structure in response to an input—was demonstrated in 1999. Nanoarchitecture, first proposed by Seeman in 1987, was beginning to be demonstrated by 2006. Also in 2006, Rothemund first demonstrated the new DNA origami technique for easily and robustly creating folded DNA molecules of any shape. In 2009, Seeman published the synthesis of a three-dimensional lattice, nearly thirty years after he had set out to do so.

Phylogenetics

In biology, phylogenetics is the study of evolutionary relatedness among various groups of organisms (for example, species or populations), which is discovered through molecular sequencing data and morphological data matrices. The term *phylogenetics* is of Greek origin from the terms *phyle/phylon*, meaning "tribe, race", and *genetikos*, meaning "relative to birth" from *genesis*. Taxonomy, the classification, identification, and naming of organisms, has been richly informed by phylogenetics but remains methodologically and logically distinct.

The fields overlap however in the science of phylogenetic systematics – often called "cladism" or "cladistics" – where only phylogenetic trees are used to delimit taxa, which represent groups of lineage-connected individuals. In biological systematics as a whole, phylogenetic analyses have become essential in researching the evolutionary tree of life.

Construction of a Phylogenetic Tree

Evolution is regarded as a branching process, whereby populations are altered over time and may speciate into separate branches, hybridize together, or terminate by extinction. This may be visualized in a phylogenetic tree.

The problem posed by phylogenetics is that genetic data are only available for living taxa, and the fossil records (osteometric data) contains less data and more-ambiguous morphological characters. A phylogenetic tree represents a hypothesis of the order in which evolutionary events are assumed to have occurred. Cladistics is the current method of choice to infer phylogenetic trees. The most commonly-used methods to infer phylogenies include parsimony, maximum likelihood, and MCMC-based Bayesian inference.

Phenetics, popular in the mid-20th century but now largely obsolete, uses distance matrix-based methods to construct trees based on overall similarity, which is often assumed to approximate phylogenetic relationships. All methods depend upon an implicit or explicit mathematical model describing the evolution of characters observed in the species included, and are usually used for molecular phylogeny, wherein the characters are aligned nucleotide or amino acid sequences.

Grouping of Organisms

There are some terms that describe the nature of a grouping in such trees. For instance, all birds and reptiles are believed to have descended from a single common ancestor, so this taxonomic grouping (yellow in the diagram below) is called monophyletic. "Modern reptile" (cyan in the diagram) is a grouping that contains a common ancestor, but does not contain all descendants of that ancestor (birds are excluded).

This is an example of a paraphyletic group. A grouping such as warm-blooded animals would include only mammals and birds (red/orange in the diagram) and is called polyphyletic because the members of this grouping do not include the most recent common ancestor.

Molecular Phylogenetics

Molecular phylogenetics, also known as molecular systematics (a term likely discouraged to avoid confusion with molecular-biological system/structure-activity relationship), is the use of the structure of molecules to gain information on an organism's evolutionary relationships. The result of a molecular phylogenetic analysis is expressed in a phylogenetic tree.

History of Molecular Phylogenetics

The theoretical frameworks for molecular systematics were laid in the 1960s in the works of Emile Zuckerkandl, Emanuel Margoliash, Linus Pauling and Walter M. Fitch. Applications of molecular systematics were pioneered by Charles G. Sibley (birds), Herbert C. Dessauer (herpetology), and Morris Goodman (primates), followed by Allan C. Wilson, Robert K. Selander, and John C. Avise (who studied various groups). Work with protein electrophoresis began around 1956. Although the results were not quantitative and did not initially improve on morphological classification, they provided tantalizing hints that long-held notions of the classifications of birds, for example, needed substantial revision. In the period of 1974–1986, DNA-DNA hybridization was the dominant technique.

Techniques and Applications

Every living organism contains DNA, RNA, and proteins. Closely related organisms generally have a high degree of agreement in the molecular structure of these substances, while the molecules of organisms distantly related usually show a pattern of dissimilarity. Conserved sequences, such as mitochondrial DNA, are expected to accumulate mutations over time, and assuming a constant rate of mutation provide a molecular clock for dating divergence. Molecular phylogeny uses such data to build a "relationship tree" that shows the probable evolution of various organisms. Not until recent decades, however, has it been possible to isolate and identify these molecular structures.

The most common approach is the comparison of homologous sequences for genes using sequence alignment techniques to identify similarity. Another application of molecular phylogeny is in DNA barcoding, where the species of an individual organism is identified using small sections of mitochondrial DNA. Another application of the techniques that make this possible can be seen in the very limited field of human genetics, such as the ever more popular use of genetic testing to determine a child's paternity, as well as the emergence of a new branch of criminal forensics focused on evidence known as genetic fingerprinting.

Theoretical Background

Early attempts at molecular systematics were also termed as chemotaxonomy and made use of proteins, enzymes, carbohydrates and other molecules which were separated and characterized using techniques such as chromatography. These have been largely replaced in recent times by DNA sequencing which produces the exact sequences of nucleotides or *bases* in either DNA or RNA segments extracted using different techniques. These are generally considered superior for evolutionary studies since the actions of evolution are ultimately reflected in the genetic sequences. At present it is still a long and expensive process to sequence the entire DNA of an organism (its genome), and this has been done for only a few species. However it is quite feasible to determine the sequence of a defined area of a particular chromosome.

Typical molecular systematic analyses require the sequencing of around 1000 base pairs. At any location within such a sequence, the bases found in a given position may vary between organisms. The particular sequence found in a given organism is referred to as its haplotype.

In principle, since there are four base types, with 1000 base pairs, we could have 4 distinct haplotypes. However, for organisms within a particular species or in a group of related species, it has been found empirically that only a minority of sites show any variation at all and most of the variations that are found are correlated, so that the number of distinct haplotypes that are found is relatively small.

In a molecular systematic analysis, the haplotypes are determined for a defined area of genetic material; ideally a substantial sample of individuals of the target species or other taxon are used however many current studies are based on single individuals. Haplotypes of individuals of closely related, but supposedly different, taxa are also determined. Finally, haplotypes from a smaller number of individuals from a definitely different taxon are determined: these are referred to as an *out group*.

The base sequences for the haplotypes are then compared. In the simplest case, the difference between two haplotypes is assessed by counting the number of locations where they have different bases: this is referred to as the number of *substitutions* (other kinds of differences between haplotypes can also occur, for example the *insertion* of a section of nucleic acid in one haplotype that is not present in another). Usually the difference between organisms is re-expressed as a *percentage divergence*, by dividing the number of substitutions by the number of base pairs analysed: the hope is that this measure will be independent of the location and length of the section of DNA that is sequenced.

An older and superseded approach was to determine the divergences between the genotypes of individuals by DNA-DNA hybridisation. The advantage claimed for using hybridisation rather than gene sequencing was that it was based on the entire genotype, rather than on particular sections of DNA. Modern sequence comparison techniques overcome this objection by the use of multiple sequences.

Once the divergences between all pairs of samples have been determined, the resulting triangular matrix of differences is submitted to some form of statistical cluster analysis, and the resulting dendrogram is examined in order to see whether the samples cluster in the way that would be expected from current ideas about the taxonomy of the group, or not. Any group of haplotypes that are all more similar to one another than any of them is to any other haplotype may be said to constitute a clade. Statistical techniques such as bootstrapping and jackknifing help in providing reliability estimates for the positions of haplotypes within the evolutionary trees.

Limitations of Molecular Systematics

Molecular systematics is an essentially cladistic approach: it assumes that classification must correspond to phylogenetic descent, and that all valid taxa must be monophyletic.

Molecular systematics often uses the molecular clock assumption that quantitative similarity of genotype is a sufficient measure of the recency of genetic divergence. Particularly in relation to speciation, this assumption could be wrong if either some genotypic modification acted to prevent interbreeding between two groups of organisms, or genetic modification proceeded at different rates in different subgroups of the organisms.

In animals, it is often convenient to use mitochondrial DNA for molecular systematic analysis. However, because in mammals mitochondria are inherited only from the mother, this is not fully satisfactory, because inheritance in the paternal line might not be detected.

Ernst Haeckel's Recapitulation Theory

During the late 19th century, Ernst Haeckel's recapitulation theory, or biogenetic law, was widely accepted. This theory was often expressed as "ontogeny recapitulates phylogeny", i.e. the development of an organism exactly mirrors the evolutionary development of the species. Haeckel's early version of this hypothesis [that the embryo mirrors *adult* evolutionary ancestors] has since been rejected, and the hypothesis amended as the embryo's development mirroring *embryos* of its evolutionary ancestors. He was accused by five professors of falsifying his images of embryos. Most modern biologists recognize numerous connections between ontogeny and phylogeny, explain them using evolutionary theory, or view them as supporting evidence for that theory. Donald I. Williamson suggested that larvae and embryos represented adults in other taxa that have been transferred by hybridization (the larval transfer theory). However, Williamson's views do not represent mainstream thought in molecular biology, and there is a significant body of evidence against the larval transfer theory.

Gene Transfer

In general, organisms can inherit genes in two ways: vertical gene transfer and horizontal gene transfer. Vertical gene transfer is the passage of genes from parent to offspring, and horizontal gene transfer or lateral gene transfer occurs when genes jump between unrelated organisms, a common phenomenon in prokaryotes.

Horizontal gene transfer has complicated the determination of phylogenies of organisms, and inconsistencies in phylogeny have been reported among specific groups of organisms depending on the genes used to construct evolutionary trees. Carl Woese came up with the three-domain theory of life (eubacteria, archaea and eukaryotes) based on his discovery that the genes encoding ribosomal RNA are ancient and distributed over all lineages of life with little or no horizontal gene transfer. Therefore, rRNAs are commonly recommended as molecular clocks for reconstructing phylogenies.

This has been particularly useful for the phylogeny of microorganisms, to which the species concept does not apply and which are too morphologically simple to be classified based on phenotypic traits.

Taxon Sampling and Phylogenetic Signal

Owing to the development of advanced sequencing techniques in molecular biology, it has become feasible to gather large amounts of data (DNA or amino acid sequences) to infer phylogenetic hypotheses. For example, it is not rare to find studies with character matrices based on whole mitochondrial genomes (~16,000 nucleotides, in many animals). However, it has been proposed that it is more important to increase the number of taxa in the matrix than to increase the number of characters, because the more taxa the more robust is the resulting phylogenetic tree.

This may be partly due to the breaking up of long branches. It has been argued that this is an important reason to incorporate data from fossils into phylogenies where possible. Of course, phylogenetic data that include fossil taxa are generally based on morphology, rather than DNA data. Using simulations, Derrick Zwickl and David Hillis found that increasing taxon sampling in phylogenetic inference has a positive effect on the accuracy of phylogenetic analyses.

Another important factor that affects the accuracy of tree reconstruction is whether the data analysed actually contain a useful phylogenetic signal, a term that is used generally to denote whether related organisms tend to resemble each other with respect to their genetic material or phenotypic traits. Ultimately, however, there is no way to measure whether a particular phylogenetic hypothesis is accurate or not, unless the "true" relationships among the taxa being examined are already known. The best result an empirical systematist can hope to attain is a tree with branches well-supported by the available evidence.

Importance of Missing Data

In general, the more data that is available when constructing a tree, the more accurate and reliable the resulting tree will be. Missing data is no less detrimental than simply having less data, although its impact is greatest when most of the missing data is in a small number of taxa. The fewer characters that have missing data, the better; concentrating the missing data across a small number of character states produces a more robust tree.

Role of Fossils

Because many morphological characters involve embryological or soft-tissue characters that cannot be fossilized, and the interpretation of fossils is more ambiguous than living taxa, it is sometimes difficult to incorporate fossil data into phylogenies. However, despite these limitations, the inclusion of fossils is invaluable, as they can provide information in sparse areas of trees, breaking up long branches and constraining intermediate character states; thus, fossil taxa contribute as much to tree resolution as modern taxa.

Molecular phylogenies can reveal rates of diversification, but in order to track rates of origination, extinction and patterns in diversification, fossil data must be incorporated. Molecular techniques assume a constant rate of diversification, which is rarely likely to be true; in some (but by no means all) cases, the assumptions inherent in interpreting the fossil record (e.g. a complete and unbiased record) are closer to being true than the assumption of a constant rate, making fossil insights more accurate than molecular reconstructions.

Homoplasy Weighting

Certain characters are more likely to be evolved convergently than others; logically, such characters should be given less weight in the reconstruction of a tree. Unfortunately the only objective way to determine convergence is by the construction of a tree – a somewhat circular method. Even so, weighting homoplasious characters does indeed lead to better-supported trees. Further refinement can be brought by weighting changes in one direction higher than changes in another; for instance, the presence of thoracic wings almost guarantees placement among the pterygote insects, although because wings are often lost secondarily, their absence does not exclude a taxon from the group.

3

Genetic Genealogy

Genetic genealogy is the application of genetics to traditional genealogy. Genetic genealogy involves the use of genealogical DNA testing to determine the level of genetic relationship between individuals.

History

The investigation of surnames in genetics can be said to go back to George Darwin, a son of Charles Darwin. In 1875, George Darwin used surnames to estimate the frequency of first-cousin marriages and calculated the expected incidence of marriage between people of the same surname (isonymy).

He arrived at a figure between 2.25% and 4.5% for cousin-marriage in the population of Great Britain, with the upper classes being on the high end and the general rural population on the low end. (His parents, Charles Darwin and Emma Wedgwood, were first cousins.) This simple study was innovative for its era.

The next stimulus toward using genetics to study family history had to wait until the 1990s, when certain locations on the Y chromosome were identified as being useful for tracing male-to-male inheritance.

Dr. Karl Skorecki, a Canadian nephrologist of Ashkenazi parentage, noticed that a Sephardic fellow-congregant who was a Kohen like himself had completely different physical features. According to Jewish tradition, all Kohanim are descended from the priest Aaron, brother of Moses. Skorecki reasoned that if Kohanim were indeed the descendants of only one man, they should have a common set of genetic markers and should perhaps preserve some family resemblance to each other.

To test that hypothesis, he contacted Professor Michael Hammer of the University of Arizona, a researcher in molecular genetics and pioneer in Y chromosome research. Their report in the *Nature* in 1997 sent shock waves through the worlds of science and religion. A particular marker was indeed more likely to be present in Jewish men from the priestly tradition than in the general Jewish population. It was apparently true that a common descent had been strictly preserved for thousands of years. Moreover, the data showed that there were very few "non-paternity events".

The first to test the new methodology in general surname research was Bryan Sykes, a molecular biologist at Oxford University. His study of the Sykes surname obtained valid results by looking at only four markers on the male chromosome. It pointed the way to genetics becoming a valuable assistant in the service of genealogy and history.

In April 2000, Family Tree DNA began offering the first genetic genealogy tests to the public. This offering marked the first time that a personal theory on the Y chromosome could be tested outside of an academic study. Additionally, Sykes' concept of a surname study, which by this time had been adopted by several other academic researchers outside of Oxford, was expanded into online Surname Projects (an early form of social network) and the effort helped spread knowledge gained through testing to interested genealogists worldwide.

In 2001, Sykes went on to write the popular book *The Seven Daughters of Eve*, which described the seven major haplogroups of European ancestors. In the wake of the book's success, and with the growing availability and affordability of genealogical DNA testing, genetic genealogy as a field began growing rapidly. By 2003, the field of DNA testing of surnames was declared officially to have "arrived" in an article by Jobling and Tyler-Smith in *Nature Reviews Genetics*. The number of firms offering tests, and the number of consumers ordering them, had risen dramatically.

Another milestone in the acceptance of genetic genealogy is the Genographic Project. The Genographic Project is a five-year research study launched in 2005 by the National Geographic Society and IBM, in partnership with the University of Arizona and Family Tree DNA. Although its goals are primarily anthropological, not genealogical, the project's sale by April 2010 of more than 350,000 of its public participation testing kits, which test the general public for either twelve STR markers on the Y chromosome or mutations on the HVR1 region of the mtDNA, has helped increase the visibility of genetic genealogy.

More state-of-the-art commercial laboratories now recommend testing at least 25 markers, since the more markers tested, the more discriminating and powerful the results will be. A 12-marker STR test is usually not discriminating enough to provide conclusive results for a common surname. Genetic laboratories such as Genebase and Family Tree DNA give the option of testing 67 Y-DNA

Markers. Annual sales of genetic genealogical tests for all companies, including the laboratories that support them, are estimated to be in the area of $60 million (2006).

Interpretation

Since the year 2000, dozens of relevant academic papers have been published, and thousands of private test results organised by surname study groups have been made available on the internet. The comparison of results may be complicated by the fact that some laboratories use different testing methods. Apparently differing results from two sources may in fact be identical, and vice-versa.

Paternal and Maternal Lineages Via DNA Testing

The two most common types of genetic genealogy tests are Y-DNA (paternal line) and mtDNA (maternal line) genealogical DNA tests. Note that Y chromosome and Y-DNA are used interchangeably in this article.

These tests involve the comparison of certain sequences of the DNA of pairs of individuals in order to estimate the probability that they share a common ancestor in a genealogical time frame and, through the use of a Bayesian model published by Bruce Walsh, to estimate the number of generations separating the two individuals from their most recent common ancestor or "mrca".

Y-DNA testing involves short tandem repeat (STR) and, sometimes, single nucleotide polymorphism (SNP) testing of the Y-chromosome. The Y-chromosome is present only in males and reveals information on the strict paternal line. These tests can provide insight into the recent (via STRs) and ancient (via SNPs) genetic ancestry. A Y-chromosome STR test will reveal a haplotype, which should be similar among all male descendants of a male ancestor. SNP tests are used to assign people to a paternal haplogroup, which defines a much larger genetic population.

mtDNA testing involves sequencing or testing the HVR-1 region, HVR-2 region or both. An mtDNA test may also include the additional SNPs needed to assign people to a maternal haplogroup—or even include the complete mtDNA. Either Y-DNA or mtDNA test results can be compared to the results of others via private or public DNA databases.

Biogeographical and Ethnic Origins

Additional DNA tests exist for determining biogeographical and ethnic origin, but these tests have less relevance for traditional genealogy.

Genetic genealogy has revealed astonishing links between peoples. For instance, it has shown that the ancient Phoenician people were ancestors of much of the present-day population of the island of Malta. Preliminary results from a study by Pierre Zalloua of the American University of Beirut and Spencer Wells, supported by a grant from National Geographic's Committee for Research and Exploration, were published in the October 2004 issue of *National Geographic*. One of the conclusions is that "more than half of the Y chromosome lineages that we see in today's Maltese population could have come in with the Phoenicians."

Human Migration

Genealogical DNA testing methods are also being used on a longer time scale to trace human migratory patterns. For example, they have been used to determine when the first humans came to North America and what path they followed. For several years, a number of researchers and laboratories from around the world have been sampling indigenous populations from around the globe in an effort to map historical human migration patterns. Recently, several projects have been created that are aimed at bringing this science to the public. One example, mentioned in History above, is the National Geographic Society's Genographic Project, which aims to map historical human migration patterns by collecting and analyzing DNA samples from over 100,000 people across five continents. Another example is the DNA Clans Genetic Ancestry Analysis, which measures a person's precise genetic connections to indigenous ethnic groups from around the world.

Typical Customers and Interest Groups

Male DNA testing customers most often start with a Y chromosome test to determine their father's paternal ancestry. Females generally begin with a mitochondrial test to trace their ancient maternal lineage, which males often have tested for the same purpose.

A common consumer goal in purchasing DNA testing services is to acquire quantified, scientific linkage to a specific ancestral group. A compelling example of this motive is found in the expressed desires of some consumers to be proven to have Viking paternal ancestry. In keeping with this marketplace demand, one British DNA testing service, Oxford Ancestors, offers a Y chromosome test purporting to assess whether given males are of "Viking stock." Those whose DNA falls into the designated haplogroup are issued Viking Descendant certificates by the testing service. The same DNA testing company participated in producing a televised documentary, "The Blood of the Vikings," in conjunction with the BBC, which showed how DNA testing could reveal Viking ancestry.

The RootsWeb Genealogy-DNA Internet discussion group has a membership of 750 subscribers from around the world. Some subscribers have had various DNA tests performed and are seeking advice and guidance in interpreting their results. The list also includes administrators of DNA projects that examine surnames, geographic regions, or ethnic groups. The sophistication of subscribers ranges from expert to novice. In some cases, subscribers have been credited with making useful and novel contributions to knowledge in the field of genetic genealogy.

Paternal and Maternal DNA Lineages

Mitochondria are small organelles that lie in the cytoplasm of eukaryotic cells, such as those of humans. Their primary purpose is to provide energy to the cell. Mitochondria are thought to be the vestigial remains of symbiotic bacteria that were once free living. One indication that mitochondria were once free living is that they contain a relatively small circular segment of DNA, called mitochondrial DNA (mtDNA). The overwhelming majority of a human's DNA is contained in chromosomes in the nucleus of the cell, but mtDNA is an exception. Individuals inherit their cytoplasm and the organelles it contains exclusively from their mothers, as these are derived from the ovum (egg cell) only, not from the sperm.

When a mutation arises in mtDNA molecule, the mutation is therefore passed in a direct female line of descent. These rare mutations are derived from copying mistakes—when the DNA is copied it is possible that a single mistake occurs in the DNA sequence, an outcome which is called a single nucleotide polymorphism (SNP).

Human Y chromosomes are male-specific sex chromosomes; nearly all humans that possess a Y chromosome will be morphologically male. Y chromosomes are therefore passed from father to son; although Y chromosomes are situated in the cell nucleus, they only recombine with the X chromosome at the ends of the Y chromosome; the vast majority of the Y chromosome (95%) does not recombine. When mutations (SNPs, and STR copying mistakes) arise in the Y chromosome, they are passed down directly from father to son in a direct male line of descent. The Y-DNA and mtDNA therefore share a certain feature: they both pass down unchanged except for mutations.

The other chromosomes, autosomes and X chromosomes in women, share their genetic material (called crossing over leading to recombination) during meiosis (a special type of cell division that occurs for the purposes of sexual reproduction). Effectively this means that the genetic material from these chromosomes gets mixed up in every generation, and so any new mutations are passed down randomly from parents to offspring. The special feature that both Y-DNA and mtDNA share, above, preserves a "written" record of their mutations because neither DNA gets mixed up or randomized—mutations remain fixed in place on both types of DNA. Furthermore the historical sequence of these mutations can also be inferred. For example, if a set of ten Y chromosomes (derived from ten different men) contains a mutation, A, but only five of these chromosomes contain a second mutation, B, it must be the case that mutation B occurred after mutation A. Furthermore all ten men who carry the chromosome with mutation A are the direct male line descendants of the same man who was the first to carry this mutation.

The first man to carry mutation B was also a direct male line descendant of this man, but is also the direct male line ancestor of all men carrying mutation B. Series of mutations such as this form molecular lineages. Furthermore each SNP mutation may define a set of specific Y chromosomes called a haplogroup.

All men carrying SNP mutation A form a single haplogroup, and all men carrying mutation B are part of this haplogroup, but mutation B (if a SNP) may also define a more recent haplogroup (which is a subgroup or subclade) of its own which men carrying only mutation A do not belong to. Both mtDNA and Y chromosomes or Y-DNA are grouped into lineages and haplogroups; these are often presented as tree-like diagrams.

Benefits

Genetic genealogy gives genealogists a means to check or supplement their genealogy results with information obtained via DNA testing. A positive test match with another individual may:

- provide locations for further genealogical research
- help determine ancestral homeland
- discover living relatives
- validate existing research
- confirm or deny suspected connections between families
- prove or disprove theories regarding ancestry.

Drawbacks

People who resist testing may cite one of the following concerns:

- Cost
- Quality of testing
- Concerns over privacy issues
- Loss of ethnic identity.

Finally, Y-DNA and mtDNA tests each only trace a single lineage (one's father's father's father's etc. lineage or one's mother's mother's mother's etc. lineage). At 10 generations back, an individual has up to 1024 unique ancestors (fewer if ancestor cousins interbred) and a Y-DNA or mtDNA test is only studying one of those ancestors, as well as their descendants and siblings (same sexed siblings for Y-DNA or all siblings for mtDNA). However, most genealogists maintain contact with many cousins (1st, 2nd, 3rd, etc., with different surnames) whose Y-DNA and mtDNA are different, and thus can be encouraged to be tested to find additional ancestral DNA lineages.

Expected Growth

Genetic genealogy is a rapidly growing field. As the cost of testing continues to drop, the number of people being tested continues to increase. The probability of finding a genetic match among the DNA databases should continue to improve. Laboratories and testing firms are engaging in active research and development that will allow for higher confidence intervals and better results interpretation, including historical interpretive reports and customized research.

Genetic Distance among Relatives

Where the genogram or family tree of individuals is known, it can be used to determine the genetic identity between individuals. It is often described as percentage of genetic identity, referring to the fraction of genome inherited from common ancestors, and not actual genomic identity, which is always approximately 99.9% identical from one human to another.

One method of calculating this genetic similarity is to do an inbreeding calculation by the path or tabular method and then multiply by 2, because any progeny would have a 1 in 2 risk of actually inheriting the identical alleles from both parents. For instance, a brother/sister relation gives 25% risk for two alleles to be identical by descent.

History of Molecular Biology

The history of molecular biology begins in the 1930s with the convergence of various, previously distinct biological disciplines: biochemistry, genetics, microbiology, and virology. With the hope of understanding life at its most fundamental level, numerous physicists and chemists also took an interest in what would become molecular biology.

In its modern sense, molecular biology attempts to explain the phenomena of life starting from the macromolecular properties that generate them. Two categories of macromolecules in particular are the focus of the molecular biologist: 1) nucleic acids, among which the most famous is deoxyribonucleic acid (or DNA), the constituent of genes, and 2) proteins, which are the active agents of living organisms.

One definition of the scope of molecular biology therefore is to characterize the structure, function and relationships between these two types of macromolecules. This relatively limited definition will suffice to allow us to establish a date for the so-called “molecular revolution”, or at least to establish a chronology of its most fundamental developments.

General Overview

In its earliest manifestations, molecular biology—the name was coined by Warren Weaver of the Rockefeller Foundation in 1938—was an ideal of physical and chemical explanations of life, rather than a coherent discipline. Following the advent of the Mendelian-chromosome theory of heredity in the 1910s and the maturation of atomic theory and quantum mechanics in the 1920s, such explanations seemed within reach. Weaver and others encouraged (and funded) research at the intersection of biology, chemistry and physics, while prominent physicists such as Niels Bohr and Erwin Schrodinger turned their attention to biological speculation.

However, in the 1930s and 1940s it was by no means clear which—if any—cross-disciplinary research would bear fruit; work in colloid chemistry, biophysics and radiation biology, crystallography, and other emerging fields all seemed promising.

In 1940, George Beadle and Edward Tatum demonstrated the existence of a precise relationship between genes and proteins. In the course of their experiments connecting genetics with biochemistry, they switched from the genetics mainstay *Drosophila* to a more appropriate model organism, the fungus *Neurospora*; the construction and exploitation of new model organisms would become a recurring theme in the development of molecular biology.

In 1944, Oswald Avery, working at the Rockefeller Institute of New York, demonstrated that genes are made up of DNA. In 1952, Alfred Hershey and Martha Chase confirmed that the genetic material of the bacteriophage, the virus which infects bacteria, is made up of DNA. In 1953, James Watson and Francis Crick discovered the double helical structure of the DNA molecule.

In 1961, Francois Jacob and Jacques Monod hypothesized the existence of an intermediary between DNA and its protein products, which they called messenger RNA. Between 1961 and 1965, the relationship between the information contained in DNA and the structure of proteins was determined: there is a code, the genetic code, which creates a correspondence between the succession of nucleotides in the DNA sequence and a series of amino acids in proteins. At the beginning of the 1960s, Monod and Jacob also demonstrated how certain specific proteins, called regulative proteins, latch onto DNA at the edges of the genes and control the transcription of these genes into messenger RNA; they direct the "expression" of the genes.

The chief discoveries of molecular biology took place in a period of only about twenty-five years. Another fifteen years were required before new and more sophisticated technologies, united today under the name of genetic engineering, would permit the isolation and characterization of genes, in particular those of highly complex organisms.

The Exploration of the Molecular Dominion

If we evaluate the molecular revolution within the context of biological history, it is easy to note that it is the culmination of a long process which began with the first observations through a microscope. The aim of these early researchers was to understand the functioning of living organisms by describing their organization at the microscopic level. From the end of the 18th century, the characterization of the chemical molecules which make up living beings gained increasingly greater attention, along with the birth of physiological chemistry in the 19th century, developed by the German chemist Justus von Liebig and following the birth of biochemistry at the beginning of the 20th, thanks to another German chemist Eduard Buchner. Between the molecules studied by chemists and the tiny structures visible under the optical microscope, such as the cellular nucleus or the chromosomes, there was an obscure zone, "the world of the ignored dimensions," as it was called by the chemical-physicist Wolfgang Ostwald. This world is populated by colloids, chemical compounds whose structure and properties were not well defined.

The successes of molecular biology derived from the exploration of that unknown world by means of the new technologies developed by chemists and physicists: X-ray diffraction, electron microscopy, ultracentrifugization, and electrophoresis. These studies revealed the structure and function of the macromolecules.

A milestone in that process was the work of Dr. Linus Pauling in 1949, which for the first time linked the specific genetic mutation in patients with sickle cell disease to a demonstrated change in an individual protein, the hemoglobin in the erythrocytes of heterozygous or homozygous individuals.

The Encounter between Biochemistry and Genetics

The development of molecular biology is also the encounter of two disciplines which made considerable progress in the course of the first thirty years of the twentieth century: biochemistry and genetics. The first studies the structure and function of the molecules which make up living things. Between 1900 and 1940, the central processes of metabolism were described: the process of digestion and the absorption of the nutritive elements derived from alimentation, such as the sugars. Every one of these processes is catalysed by a particular enzyme. Enzymes are proteins, like the antibodies present in blood or the proteins responsible for muscular contraction. As a consequence, the study of proteins, of their structure and synthesis, became one of the principal objectives of biochemists.

The second discipline of biology which developed at the beginning of the 20th century is genetics. After the rediscovery of the laws of Mendel through the studies of Hugo de Vries, Carl Correns and Erich von Tschermak in 1900, this science began to take shape thanks to the adoption by Thomas Hunt Morgan, in 1910, of a model organism for genetic studies, the famous fruit fly (Drosophila melanogaster). Shortly after, Morgan showed that the genes are localized on chromosomes.

Following this discovery, he continued working with Drosophila and, along with numerous other research groups, confirmed the importance of the gene in the life and development of organisms. Nevertheless, the chemical nature of genes and their mechanisms of action remained a mystery. Molecular biologists committed themselves to the determination of the structure, and the description of the complex relations between, genes and proteins.

The development of molecular biology was not just the fruit of some sort of intrinsic "necessity" in the history of ideas, but was a characteristically historical phenomenon, with all of its unknowns, imponderables and contingencies: the remarkable developments in physics at the beginning of the 20th century highlighted the relative lateness in development in biology, which became the "new frontier" in the search for knowledge about the empirical world.

Moreover, the developments of the theory of information and cybernetics in the 1940s, in response to military exigencies, brought to the new biology a significant number of fertile ideas and, especially, metaphors.

The choice of bacteria and of its virus, the bacteriophage, as models for the study of the fundamental mechanisms of life was almost natural - they are the smallest living organisms known to exist - and at the same time the fruit of individual choices. This model owes its success, above all, to the fame and the sense of organization of Max Delbruck, a German physicist, who was able to create a dynamic research group, based in the United States, whose exclusive scope was the study of the bacteriophage: the *School of the Phage*.

The geographic panorama of the developments of the new biology was conditioned above all by preceding work. The US, where genetics had developed the most rapidly, and the UK, where there was a coexistence of both genetics and biochemical research of highly advanced levels, were in the avant-garde. Germany, the cradle of the revolutions in physics, with the best minds and the most advanced laboratories of genetics in the world, should have had a primary role in the development of molecular biology.

But history decided differently: the arrival of the Nazis in 1933 - and, to a less extreme degree, the rigidification of totalitarian measures in fascist Italy - caused the emigration of a large number of Jewish and non-Jewish scientists. The majority of them fled to the US or the UK, providing an extra impulse to the scientific dynamism of those nations. These movements ultimately made molecular biology a truly international science from the very beginnings.

History of DNA Biochemistry

The study of DNA is a central part of molecular biology.

First Isolation of DNA

Working in the 19th century, biochemists initially isolated DNA and RNA (mixed together) from cell nuclei. They were relatively quick to appreciate the polymeric nature of their "nucleic acid" isolates, but realized only later that nucleotides were of two types—one containing ribose and the other deoxyribose. It was this subsequent discovery that led to the identification and naming of DNA as a substance distinct from RNA.

Friedrich Miescher (1844–1895) discovered a substance he called "nuclein" in 1869. Somewhat later, he isolated a pure sample of the material now known as DNA from the sperm of salmon, and in 1889 his pupil, Richard Altmann, named it "nucleic acid". This substance was found to exist only in the chromosomes.

In 1919 Phoebus Levene at the Rockefeller Institute identified the components (the four bases, the sugar and the phosphate chain) and he showed that the components of DNA were linked in the order phosphate-sugar-base. He called each of these units a nucleotide and suggested the DNA molecule consisted of a string of nucleotide units linked together through the phosphate groups, which are the 'backbone' of the molecule. However Levene thought the chain was short and that the bases repeated in the same fixed order. Torbjorn Caspersson and Einar Hammersten showed that DNA was a polymer.

Chromosomes and Inherited Traits

Max Delbruck, Nikolai V. Timofeeff-Ressovsky, and Karl G. Zimmer published results in 1935 suggesting that chromosomes are very large molecules the structure of which can be changed by treatment with X-rays, and that by so changing their structure it was possible to change the heritable characteristics governed by those chromosomes. In 1937 William Astbury produced the first X-ray diffraction patterns from DNA. He was not able to propose the correct structure but the patterns showed that DNA had a regular structure and therefore it might be possible to deduce what this structure was.

In 1943, Oswald Theodore Avery and a team of scientists discovered that traits proper to the "smooth" form of the *Pneumococcus* could be transferred to the "rough" form of the same bacteria merely by making the killed "smooth" (S) form available to the live "rough" (R) form. Quite unexpectedly, the living R *Pneumococcus* bacteria were transformed into a new strain of the S form, and the transferred S characteristics turned out to be heritable. Avery called the medium of transfer of traits the transforming principle; he identified DNA as the transforming principle, and not protein as previously thought. He essentially redid Frederick Griffith's experiment. In 1953, Alfred Hershey and Martha Chase did an experiment (Hershey-Chase experiment) that showed, in T2 phage, that DNA is the genetic material (Hershey shared the Nobel prize with Luria).

Discovery of the Structure of DNA

In the 1950s, three groups made it their goal to determine the structure of DNA. The first group to start was at King's College London and was led by Maurice Wilkins and was later joined by Rosalind Franklin. Another group consisting of Francis Crick and James D. Watson was at Cambridge. A third group was at Caltech and was led by Linus Pauling. Crick and Watson built physical models using metal rods and balls, in which they incorporated the known chemical structures of the nucleotides, as well as the known position of the linkages joining one nucleotide to the next along the polymer. At King's College Maurice Wilkins and Rosalind Franklin examined X-ray diffraction patterns of DNA fibres. Of the three groups, only the London group was able to produce good quality diffraction patterns and thus produce sufficient quantitative data about the structure.

Helix Structure

In 1948 Pauling discovered that many proteins included helical shapes. Pauling had deduced this structure from X-ray patterns and from attempts to physically model the structures. (Pauling was also later to suggest an incorrect three chain helical DNA structure based on Astbury's data.) Even in the initial diffraction data from DNA by Maurice Wilkins, it was evident that the structure involved helices. But this insight was only a beginning. There remained the questions of how many strands came together, whether this number was the same for every helix, whether the bases pointed toward the helical axis or away, and ultimately what were the explicit angles and coordinates of all the bonds and atoms. Such questions motivated the modelling efforts of Watson and Crick.

Complementary Nucleotides

In their modelling, Watson and Crick restricted themselves to what they saw as chemically and biologically reasonable. Still, the breadth of possibilities was very wide. A breakthrough occurred in 1952, when Erwin Chargaff visited Cambridge and inspired Crick with a description of experiments Chargaff had published in 1947. Chargaff had observed that the proportions of the four nucleotides vary between one DNA sample and the next, but that for particular pairs of nucleotides — adenine and thymine, guanine and cytosine — the two nucleotides are always present in equal proportions. Using X-ray diffraction, as well as other data from Rosalind Franklin and her information that the bases were paired, James D. Watson and Francis Crick arrived at the first accurate model of DNA's molecular structure in 1953, which was accepted through inspection by Rosalind Franklin.

The discovery was made on February 28, 1953; the first Watson/Crick paper appeared in Nature on April 25, 1953. Sir Lawrence Bragg, the director of the Cavendish Laboratory, where Watson and Crick worked, gave a talk at Guys Hospital Medical School in London on Thursday, May 14, 1953 which resulted in an article by Ritchie Calder in The News Chronicle of London, on Friday, May 15, 1953, entitled "Why You Are You. Nearer Secret of Life." The news reached readers of The New York Times the next day; Victor K. McElheny, in researching his biography, "Watson and DNA: Making a Scientific Revolution", found a clipping of a six-paragraph New York Times article written from London and dated May 16, 1953 with the headline "Form of `Life Unit' in Cell Is Scanned." The article ran in an early edition and was then pulled to make space for news deemed more important.(The New York Times subsequently ran a longer article on June 12, 1953). The Cambridge University undergraduate newspaper also ran its own short article on the discovery on Saturday, May 30, 1953. Bragg's original announcement at a Solvay conference on proteins in Belgium on 8 April 1953 went unreported by the press! In 1962 Watson, Crick, and Maurice Wilkins jointly received the Nobel Prize for Physiology or Medicine for their determination of the structure of DNA.

"Central Dogma"

Watson and Crick's model attracted great interest immediately upon its presentation. Arriving at their conclusion on February 21, 1953, Watson and Crick made their first announcement on February 28. In an influential presentation in 1957, Crick laid out the "Central Dogma", which foretold the relationship between DNA, RNA, and proteins, and articulated the "sequence hypothesis." A critical confirmation of the replication mechanism that was implied by the double-helical structure followed in 1958 in the form of the Meselson-Stahl experiment. Work by Crick and coworkers showed that the genetic code was based on non-overlapping triplets of bases, called codons, and Har Gobind Khorana and others deciphered the genetic code not long afterward (1966). These findings represent the birth of molecular biology.

History of RNA Tertiary Structure

Pre-history : The Helical Structure of RNA

The earliest work in RNA structural biology coincided, more or less, with the work being done on DNA in the early 1950s. In their seminal 1953 paper, Watson and Crick suggested that van der Waals crowding by the 2'OH group of ribose would preclude RNA from adopting a double helical structure identical to the model they proposed - what we now know as B-form DNA. This provoked questions about the three dimensional structure of RNA: could this molecule form some type of helical structure, and if so, how? As with DNA, early structural work on RNA centered around isolation of native RNA polymers for fibre diffraction analysis. In part because of heterogeneity of the samples tested, early fibre diffraction patterns were usually ambiguous and not readily interpretable. In 1955, Grunberg-Manago *et al.* published a paper describing the enzyme polynucleotide phosphorylase, which cleaved a phosphate group from nucleotide diphosphates to catalyse their polymerization.

This discovery allowed researchers to synthesize homogenous nucleotide polymers, which they then combined to produce double stranded molecules. These samples yielded the most readily interpretable fibre diffraction patterns yet obtained, suggesting an ordered, helical structure for cognate, double stranded RNA that differed from that observed in DNA. These results paved the way for a series of investigations into the various properties and propensities of RNA. Through the late 1950s and early 1960s, numerous papers were publihed on various topics in RNA structure, including RNA-DNA hybridization, triple stranded RNA, and even small-scale crystallography of RNA di-nucleotides - G-C, and A-U - in primitive helix-like arrangements. For a more in-depth review of the early work in RNA structural biology.

The Beginning: Crystal Structure of tRNA

In the mid-1960s, the role of tRNA in protein synthesis was being intensively studied. At this point, ribosomes had been implicated in protein synthesis, and it had been shown that an mRNA strand was necessary for the formation of these structures. In a 1964 publication, Warner and Rich showed that ribosomes active in protein synthesis contained tRNA molecules bound at the A and P sites, and discussed the notion that these molecules aided in the peptidyl transferase reaction.

However, despite considerable biochemical characterization, the structural basis of tRNA function remained a mystery. In 1965, Holley *et al.* purified and sequenced the first tRNA molecule, initially proposing that it adopted a cloverleaf structure, based largely on the ability of certain regions of the molecule to form stem loop structures. The isolation of tRNA proved to be the first major windfall in RNA structural biology. Following Holley's publication, numerous investigators began work on isolation tRNA for crystallographic study, developing improved methods for isolating the molecule as they worked. By 1968 several groups had produced tRNA crystals, but these proved to be of limited quality and did not yield data at the resolutions necessary to determine structure. In 1971, Kim *et al.* achieved another breakthrough, producing crystals of yeast tRNA that diffracted to 2-3 Angstrom resolutions by using spermine, a naturally occurring polyamine, which bound to and stabilized the tRNA. Despite having suitable crystals, however, the structure of tRNA was not immediately solved at high resolution; rather it took pioneering work in the use of heavy metal derivatives and a good deal more time to produce a high-quality density map of the entire molecule. In 1973, Kim *et al.* produced a 4 Angstrom map of the tRNA molecule in which they could unambiguously trace the entire backbone. This solution would be followed by many more, as various investigators worked to refine the structure and thereby more thoroughly elucidate the details of base pairing and stacking interactions, and validate the published architecture of the molecule.

The tRNA structure is notable in the field of nucleic acid structure in general, as it represented the first solution of a long-chain nucleic acid structure of any kind - RNA or DNA - preceding Dickerson's solution of a B-form dodecamer by nearly a decade. Also, tRNA demonstrated many of the tertiary interactions observed in RNA architecture which would not be categorized and more thoroughly understood for years to come, providing a foundation for all future RNA structural research.

The Renaissance: The Hammerhead Ribozyme and the Group I Intron: $P_{4\text{-}6}$

For a considerable time following the first tRNA structures, the field of RNA structure did not dramatically advance. The ability to study an RNA strcuture depended upon the potential to isolate the RNA target. This proved limiting to the field for many years, in part owing to the fact that other know targets - i.e. the ribosome - were significantly more difficult to isolate and crystallize. Further, because other interesting RNA targets had simply not been identified, or were not sufficiently understood to be deemed interesting, there was simply a lack of things to study structurally. As such, for some twenty years following the original publication of the tRNA structure, the structures of only a handful of other RNA targets were solved, with almost all of these belonging to the transfer RNA family. This unfortunate lack of scope would eventually be overcome largely because of two major advancements in nucleic acid research: the identification of ribozymes, and the ability to produce them via *in vitro* transcription.

Subsequent to Tom Cech's publication implicating the *Tetrahymena* group I intron as an autocatalytic ribozyme, and Sidney Altman's report of catalysis by ribonuclease P RNA, several other catalytic RNAs were identified in the late 1980s, including the hammerhead ribozyme. In 1994, McKay *et al.* published the structure of a 'hammerhead RNA-DNA ribozyme-inhibitor complex' at 2.6 Angstrom resolution, in which the autocatalytic activity of the ribozyme was disrupted via binding to a DNA substrate. The conformation of the ribozyme published in this paper was eventually shown to be one of several possible states, and although this particular sample was catalytically inactive, subsequent structures have revealed its active-state architecture. This structure was followed by Doudna's publication of the structure of the P4-P6 domains of the *Tetrahymena* group I intron, a fragment of the ribozyme originally made famous by Cech. The second clause in the title of this publication - *Principles of RNA Packing* - concisely evinces the value of these two structures: for the first time, comparisons could be made between well described tRNA structures and those of globular RNAs outside the transfer family. This allowed the framework of categorization to be built for RNA tertiary structure. It was now possible to propose the conservation of motifs, folds, and various local stabilizing interactions. For an early review of these structures and their implications.

In addition to the advances being made in global structure determination via crystallography, the early 1990s also saw the implementation of NMR as a powerful technique in RNA structural biology. Coincident with the large-scale ribozyme structures being solved crystallographically, a number of structures of small RNAs and RNAs complexed with drugs and peptides were solved using NMR. In addition, NMR was now being used to investigate and supplement crystal structures, as exemplified by the determination of an isolated tetraloop-receptor motif structure published in 1997.

Investigations such as this enabled a more precise characterization of the base pairing and base stacking interactions which stabilized the global folds of large RNA molecules. The importance of understanding RNA tertiary structural motifs was prophetically well described by Michel and Costa in their publication identifying the tetraloop motif: "..it should not come as a surprise if self-folding RNA molecules were to make intensive use of only a relatively small set of tertiary motifs. Identifying these motifs would greatly aid modelling enterprises, which will remain essential as long as the crystallization of large RNAs remains a difficult task".

The Modern Era: The Age of RNA Structural Biology

The resurgence of RNA structural biology in the mid-1990s has caused a veritable explosion in the field of nucleic acid structural research. Since the publication of the hammerhead and $P_{4\text{-}6}$ structures, numerous major contributions to the field have been made. Some of the most noteworthy examples include the structures of the Group I and Group II introns, and the Ribosome.

It should be noted that the first three structures were produced using *in vitro* transcription, and that NMR has played a role in investigating partial components of all four structures - testaments to the indispensability of both techniques for RNA research. Most recently, the 2009 Nobel Prize in Chemistry was awarded to Ada Yonath, Venkatraman Ramakrishnan, and Thomas Steitz for their structural work on the ribosome, demonstrating the prominent role RNA structural biology has taken in modern molecular biology.

History of Protein Biochemistry

First Isolation and Classification

Proteins were recognized as a distinct class of biological molecules in the eighteenth century by Antoine Fourcroy and others. Members of this class (called the "albuminoids", *Eiweisskorper*, or *matières albuminoides*) were recognized by their ability to coagulate or flocculate under various treatments such as heat or acid; well-known examples at the start of the nineteenth century included albumen from egg whites, blood serum albumin, fibrin, and wheat gluten. The similarity between the cooking of egg whites and the curdling of milk was recognized even in ancient times; for example, the name *albumen* for the egg-white protein was coined by Pliny the Elder from the Latin *albus ovi* (egg white).

With the advice of Jons Jakob Berzelius, the Dutch chemist Gerhardus Johannes Mulder carried out elemental analyses of common animal and plant proteins. To everyone's surprise, all proteins had nearly the same empirical formula, roughly $C_{400}H_{620}N_{100}O_{120}$ with individual sulfur and phosphorus atoms. Mulder published his findings in two papers (1837,1838) and hypothesized that there was one basic substance (*Grundstoff*) of proteins, and that it was synthesized by plants and absorbed from them by animals in digestion. Berzelius was an early proponent of this theory and proposed the name "protein" for this substance in a letter dated 10 July 1838

Mulder went on to identify the products of protein degradation such as the amino acid, leucine, for which he found a (nearly correct) molecular weight of 131 Da.

Purifications and Measurements of Mass

The minimum molecular weight suggested by Mulder's analyses was roughly 9 kDa, hundreds of times larger than other molecules being studied. Hence, the chemical structure of proteins (their primary structure) was an active area of research until 1949, when Fred Sanger sequenced insulin.

The (correct) theory that proteins were linear polymers of amino acids linked by peptide bonds was proposed independently and simultaneously by Franz Hofmeister and Emil Fischer at the same conference in 1902. However, some scientists were sceptical that such long macromolecules could be stable in solution. Consequently, numerous alternative theories of the protein primary structure were proposed, e.g., the colloidal hypothesis that proteins were assemblies of small molecules, the cyclol hypothesis of Dorothy Wrinch, the diketopiperazine hypothesis of Emil Abderhalden and the pyrrol/piperidine hypothesis of Troensgard (1942). Most of these theories had difficulties in accounting for the fact that the digestion of proteins yielded peptides and amino acids. Proteins were finally shown to be macromolecules of well-defined composition (and not colloidal mixtures) by Theodor Svedberg using analytical ultracentrifugation. The possibility that some proteins are non-covalent associations of such macromolecules was shown by Gilbert Smithson Adair (by measuring the osmotic pressure of hemoglobin) and, later, by Frederic M. Richards in his studies of ribonuclease S. The mass spectrometry of proteins has long been a useful technique for identifying posttranslational modifications and, more recently, for probing protein structure.

Most proteins are difficult to purify in more than milligram quantities, even using the most modern methods. Hence, early studies focused on proteins that could be purified in large quantities, e.g., those of blood, egg white, various toxins, and digestive/metabolic enzymes obtained from slaughterhouses. Many techniques of protein purification were developed during World War II in a project led by Edwin Joseph Cohn to purify blood proteins to help keep soldiers alive. In the late 1950s, the Armour Hot Dog Co. purified 1 kg (= one million milligrams) of pure bovine pancreatic ribonuclease A and made it freely available to scientists around the world. This generous act made RNase A the main protein for basic research for the next few decades, resulting in several Nobel Prizes.

Protein Folding and First Structural Models

The study of protein folding began in 1910 with a famous paper by Henrietta Chick and C. J. Martin, in which they showed that the flocculation of a protein was composed of two distinct processes: the precipitation of a protein from solution was *preceded* by another process called denaturation, in which the protein became much less soluble, lost its enzymatic activity and became more chemically reactive. In the mid-1920s, Tim Anson and Alfred Mirsky proposed that denaturation was a reversible process, a correct hypothesis that was initially lampooned by some scientists as “unboiling the egg”. Anson also suggested that denaturation was a two-state (“all-or-none”) process, in which one fundamental molecular transition resulted in the drastic changes in solubility, enzymatic activity and chemical reactivity; he further noted that the free energy changes upon denaturation were much smaller than those typically involved in chemical reactions.

In 1929, Hsien Wu hypothesized that denaturation was protein unfolding, a purely conformational change that resulted in the exposure of amino acid side chains to the solvent. According to this (correct) hypothesis, exposure of aliphatic and reactive side chains to solvent rendered the protein less soluble and more reactive, whereas the loss of a specific conformation caused the loss of enzymatic activity. Although considered plausible, Wu’s hypothesis was not immediately accepted, since so little was known of protein structure and enzymology and other factors could account for the changes in solubility, enzymatic activity and chemical reactivity. In the early 1960s, Chris Anfinsen showed that the folding of ribonuclease A was fully reversible with no external cofactors needed, verifying the "thermodynamic hypothesis" of protein folding that the folded state represents the global minimum of free energy for the protein.

The hypothesis of protein folding was followed by research into the physical interactions that stabilize folded protein structures. The crucial role of hydrophobic interactions was hypothesized by Dorothy Wrinch and Irving Langmuir, as a mechanism that might stabilize her cyclol structures. Although supported by J. D. Bernal and others, this (correct) hypothesis was rejected along with the cyclol hypothesis, which was disproven in the 1930s by Linus Pauling (among others). Instead, Pauling championed the idea that protein structure was stabilized mainly by hydrogen bonds, an idea advanced initially by William Astbury (1933). Remarkably, Pauling's incorrect theory about H-bonds resulted in his *correct* models for the secondary structure elements of proteins, the alpha helix and the beta sheet.

The hydrophobic interaction was restored to its correct prominence by a famous article in 1959 by Walter Kauzmann on denaturation, based partly on work by Kaj Linderstrom-Lang. The ionic nature of proteins was demonstrated by Bjerrum, Weber and Arne Tiselius, but Linderstrom-Lang showed that the charges were generally accessible to solvent and not bound to each other (1949).

The secondary and low-resolution tertiary structure of globular proteins was investigated initially by hydrodynamic methods, such as analytical ultracentrifugation and flow birefringence. Spectroscopic methods to probe protein structure (such as circular dichroism, fluorescence, near-ultraviolet and infrared absorbance) were developed in the 1950s. The first atomic-resolution structures of proteins were solved by X-ray crystallography in the 1960s and by NMR in the 1980s. As of 2006, the Protein Data Bank has nearly 40,000 atomic-resolution structures of proteins. In more recent times, cryo-electron microscopy of large macromolecular assemblies and computational protein structure prediction of small protein domains are two methods approaching atomic resolution.

RNA

Ribonucleic acid (RNA) is one of the three major macromolecules (along ith DNA and proteins) that are essential for all known forms of life. Like DNA, RNA is made up of a long chain of components called nucleotides. Each nucleotide consists of a nucleobase (sometimes called a nitrogenous base), a ribose sugar, and a phosphategroup. The sequence of nucleotides allows RNA to encode genetic information. For example, someviruses use RNA instead of DNA as their genetic material, and all organisms use messenger RNA(mRNA) to carry the genetic information that directs the synthesis of proteins.

Like proteins, some RNA molecules play an active role in cells by catalyzing biological reactions, controlling gene expression, or sensing and communicating responses to cellular signals. One of these active processes is protein synthesis, a universal function whereby mRNA molecules direct the assembly of proteins on ribosomes. This process uses transfer RNA (tRNA) molecules to deliver amino acids to the ribosome, where ribosomal RNA (rRNA) links amino acids together to form proteins.

The chemical structure of RNA is very similar to that of DNA, with two differences—(a) RNA contains the sugar ribose while DNA contains the slightly different sugar deoxyribose (a type of ribose that lacks one oxygen atom), and (b) RNA has the nucleobase uracil while DNA contains thymine (uracil and thymine have similar base-pairing properties).

Unlike DNA, most RNA molecules are single-stranded. Single-stranded RNA molecules adopt very complex three-dimensional structures, since they are not restricted to the repetitive double-helical form of double-stranded DNA. RNA is made within living cells by RNA polymerases, enzymes which act to copy a DNA or RNA template into a new RNA strand through processes known as transcriptionor RNA replication, respectively.

Comparison with DNA

RNA and DNA are both nucleic acids, but differ in three main ways. First, unlike DNA which is generally double-stranded, RNA is a single-stranded molecule in many of its biological roles and has a much shorter chain of nucleotides. Second, while DNA contains *deoxyribose*, RNA contains *ribose* (there is no hydroxyl group attached to the pentose ring in the 2' position in DNA). These hydroxyl groups make RNA less stable than DNA because it is more prone to hydrolysis. Third, the complementary base to adenine is not thymine, as it is in DNA, but rather uracil, which is anunmethylated form of thymine.

Like DNA, most biologically active RNAs, including mRNA, tRNA, rRNA, snRNAs and other non-coding RNAs, contain self-complementary sequences that allow parts of the RNA to fold and pair with itself to form double helices. Structural analysis of these RNAs has revealed that they are highly structured. Unlike DNA, their structures do not consist of long double helices but rather collections of short helices packed together into structures akin to proteins. In this fashion, RNAs can achieve chemical catalysis, like enzymes. For instance, determination of the structure of the ribosome—an enzyme that catalyses peptide bond formation—revealed that its active site is composed entirely of RNA.

Structure

Each nucleotide in RNA contains a ribose sugar, with carbons numbered 1' through 5'. A base is attached to the 1' position, generally adenine (A), cytosine (C), guanine (G) or uracil (U). Adenine and guanine are purines, cytosine and uracil are pyrimidines. A phosphate group is attached to the 3' position of one ribose and the 5' position of the next. The phosphate groups have a negative charge each at physiological pH, making RNA a charged molecule (polyanion).

The bases may form hydrogen bonds between cytosine and guanine, between adenine and uracil and between guanine and uracil. However other interactions are possible, such as a group of adenine bases binding to each other in a bulge, or the GNRA tetraloop that has a guanine–adenine base-pair.n important structural feature of RNA that distinguishes it from DNA is the presence of ahydroxyl group at the 2' position of the ribose sugar. The presence of this functional group causes the helix to adopt the A-form geometry rather than the B-form most commonly observed in DNA.

This results in a very deep and narrow major groove and a shallow and wide minor groove. A second consequence of the presence of the 2'-hydroxyl group is that in conformationally flexible regions of an RNA molecule (that is, not involved in formation of a double helix), it can chemically attack the adjacent phosphodiester bond to cleave the backbone. RNA is transcribed with only four bases (adenine, cytosine, guanine and uracil), but these bases and attached sugars can be modified in numerous ways as the RNAs mature. Pseudouridine (ψ), in which the linkage between uracil and ribose is changed from a C–N bond to a C–C bond, and ribothymidine (T), are found in various places (most notably in the TψC loop of tRNA). Another notable modified base is hypoxanthine, a deaminated adenine base whose nucleoside is called inosine (I). Inosine plays a key role in the wobble hypothesis of thegenetic code.

There are nearly 100 other naturally occurring modified nucleosides, of which pseudouridine and nucleosides with 2'-O-methylribose are the most common. The specific roles of many of these modifications in RNA are not fully understood. However, it is notable that in ribosomal RNA, many of the post-transcriptional modifications occur in highly functional regions, such as the peptidyl transferase centre and the subunit interface, implying that they are important for normal function. The functional form of single stranded RNA molecules, just like proteins, frequently requires a specific tertiary structure. The scaffold for this structure is provided by secondary structural elements which are hydrogen bonds within the molecule. This leads to several recognizable "domains" of secondary structure like hairpin loops, bulges and internal loops. Since RNA is charged, metal ions such as Mg^{2+} are needed to stabilise many secondary and tertiary structures.

Synthesis

Synthesis of RNA is usually catalysed by an enzyme—RNA polymerase—using DNA as a template, a process known as transcription. Initiation of transcription begins with the binding of the enzyme to a promoter sequence in the DNA (usually found "upstream" of a gene). The DNA double helix is unwound by the helicase activity of the enzyme. The enzyme then progresses along the template strand in the 3' to 5' direction, synthesizing a complementary RNA molecule with elongation occurring in the 5' to 3' direction. The DNA sequence also dictates where termination of RNA synthesis will occur.

RNAs are often modified by enzymes after transcription. For example, a poly(A) tail and a 5' cap are added to eukaryotic pre-mRNA andintrons are removed by the spliceosome. There are also a number of RNA-dependent RNA polymerases that use RNA as their template for synthesis of a new strand of RNA. For instance, a number of RNA viruses (such as poliovirus) use this type of enzyme to replicate their genetic material. Also, RNA-dependent RNA polymerase is part of the RNA interference pathway in many organisms.

Messenger RNA

Messenger RNA (mRNA) is a molecule of RNA encoding a chemical "blueprint" for a protein product. mRNA is transcribed from a DNA template, and carries coding information to the sites of protein synthesis: the ribosomes. Here, the nucleic acid polymer is translated into a polymer of amino acids: a protein. In mRNA as in DNA, genetic information is encoded in the sequence of nucleotides arranged into codonsconsisting of three bases each. Each codon encodes for a specific amino acid, except the stop codons that terminate protein synthesis. This process requires two other types of RNA: transfer RNA (tRNA) mediates recognition of the codon and provides the corresponding amino acid, while ribosomal RNA (rRNA) is the central component of the ribosome's protein manufacturing machinery.

Synthesis, Processing, and Function

The brief existence of an mRNA molecule begins with transcription and ultimately ends in degradation. During its life, an mRNA molecule may also be processed, edited, and transported prior to translation. Eukaryotic mRNA molecules often require extensive processing and transport, while prokaryotic molecules do not.

Transcription

During transcription, RNA polymerase makes a copy of a gene from the DNA to mRNA as needed. This process is similar in eukaryotes and prokaryotes. One notable difference, however, is that prokaryotic RNA polymerase associates with mRNA processing enzymes during transcription so that processing can proceed quickly after the start of transcription. The short-lived, unprocessed or partially processed, product is termed *pre-mRNA*; once completely processed, it is termed *mature mRNA*.

Eukaryotic Pre-mRNA Processing

Processing of mRNA differs greatly among eukaryotes, bacteria and archea. Non-eukaryotic mRNA is essentially mature upon transcription and requires no processing, except in rare cases. Eukaryotic pre-mRNA, however, requires extensive processing.

5' Cap Addition

A *5' cap* (also termed an RNA cap, an RNA 7-methylguanosine cap or an RNA m^7G cap) is a modified guanine nucleotide that has been added to the "front" or 5' end of a eukaryotic messenger RNA shortly after the start of transcription. The 5' cap consists of a terminal 7-methylguanosine residue which is linked through a 5'-5'-triphosphate bond to the first transcribed nucleotide. Its presence is critical for recognition by the ribosome and protection from RNases.

Cap addition is coupled to transcription, and occurs co-transcriptionally, such that each influences the other. Shortly after the start of transcription, the 5' end of the mRNA being synthesized is bound by a cap-synthesizing complex associated with RNA polymerase. Thisenzymatic complex catalyses the chemical reactions that are required for mRNA capping. Synthesis proceeds as a multi-step biochemicalreaction.

Splicing

Splicing is the process by which pre-mRNA is modified to remove certain stretches of non-coding sequences called introns; the stretches that remain include protein-coding sequences and are called exons. Sometimes pre-mRNA messages may be spliced in several different ways, allowing a single gene to encode multiple proteins. This process is called alternative splicing. Splicing is usually performed by an RNA-protein complex called the spliceosome, but some RNA molecules are also capable of catalyzing their own splicing.

Editing

In some instances, an mRNA will be edited, changing the nucleotide composition of that mRNA. An example in humans is the apolipoprotein B mRNA, which is edited in some tissues, but not others. The editing creates an early stop codon, which upon translation, produces a shorter protein.

Polyadenylation

Polyadenylation is the covalent linkage of a polyadenylyl moiety to a messenger RNA molecule. In eukaryotic organisms, most messenger RNA (mRNA) molecules are polyadenylated at the 3' end. The poly(A) tail and the protein bound to it aid in protecting mRNA from degradation by exonucleases. Polyadenylation is also important for transcription termination, export of the mRNA from the nucleus, and translation. mRNA can also be polyadenylated in prokaryotic organisms, where poly(A) tails act to facilitate, rather than impede, exonucleolytic degradation.

Polyadenylation occurs during and immediately after transcription of DNA into RNA. After transcription has been terminated, the mRNA chain is cleaved through the action of an endonuclease complex associated with RNA polymerase. After the mRNA has been cleaved, around 250 adenosine residues are added to the free 3' end at the cleavage site.

This reaction is catalysed by polyadenylate polymerase. Just as in alternative splicing, there can be more than one polyadenylation variant of a mRNA.

Transport

Another difference between eukaryotes and prokaryotes is mRNA transport. Because eukaryotic transcription and translation is compartmentally separated, eukaryotic mRNAs must be exported from the nucleus to the cytoplasm. Mature mRNAs are recognized by their processed modifications and then exported through the nuclear pore.

In neurons mRNA must be transported from the soma to the dendriteswhere local translation occurs in response to external stimuli. Many messages are marked with so-called "zip codes" which targets their transport to a specific location.

Translation

Because prokaryotic mRNA does not need to be processed or transported, translation by the ribosome can begin immediately after the end of transcription. Therefore, it can be said that prokaryotic translation is *coupled* to transcription and occurs *co-transcriptionally*. Eukaryotic mRNA that has been processed and transported to the cytoplasm (i.e. mature mRNA) can then be translated by the ribosome. Translation may occur at ribosomes free-floating in the cytoplasm, or directed to the endoplasmic reticulum by the signal recognition particle. Therefore, unlike prokaryotes, eukaryotic translation *is not* directly coupled to transcription.

5' cap

The 5' cap is a specially altered nucleotide on the 5' end of precursor messenger RNA and some other primary RNA transcripts as found ineukaryotes and, as a special exception, caliciviruses such as noroviruses. The process of 5' capping is vital to creating mature messenger RNA, which is then able to undergo translation. Capping ensures the messenger RNA's stability while it undergoes translation in the process of protein synthesis, and is a highly regulated process that occurs in the cell nucleus.

5' Cap Structure

The 5' cap is found on the 5' end of an mRNA molecule and consists of a guanine nucleotide connected to the mRNA via an unusual 5' to 5' triphosphate linkage. This guanosine is methylated on the 7 position directly after capping *in vitro* by a methyl transferase. It is referred to as a 7-methylguanosine cap, abbreviated m^7G.

Further modifications include the possible methylation of the 2' hydroxy-groups of the first 3 ribose sugars of the 5' end of the mRNA. The methylation of both 2' hydroxy-groups is shown on the diagram. The 5' cap looks like the 3' end of an RNA molecule (the 5' carbon of the cap ribose is bonded, and the 3' unbonded). This provides significant resistance to 5' exonucleases.

Capping Process

The starting point is the unaltered 5' end of an RNA molecule. This features a final nucleotide followed by three phosphate groups attached to the 5' carbon.

1. One of the terminal phosphate groups is removed (by a phosphatase), leaving two terminal phosphates.
2. GTP is added to the terminal phosphates (by a guanylyl transferase), losing two phosphate groups (from the GTP) in the process. This results in the 5' to 5' triphosphate linkage.
3. The 7-Nitrogen of guanine is methylated (by a methyl transferase).
4. Other methyltransferases are optionally used to carry out methylation of 5' proximal nucleotides.

5' Capping Targeting

The Capping Enzyme Complex (CEC) required for capping is found bound to the RNA polymerase II before transcription starts. As soon as the 5' end of the new transcript emerges the enzymes transfer to it and begin the capping process (this is a similar kind of mechanism to ensure capping as for polyadenylation).

The enzymes for capping can only bind to RNA polymerase II ensuring specificity to only these transcripts, which are almost entirely mRNA.

5' Cap Function

The 5' cap has 4 main functions:

1. Regulation of nuclear export.
2. Prevention of degradation by exonucleases.
3. Promotion of translation.
4. Promotion of 5' proximal intron excision.

Nuclear export of RNA is regulated by the Cap binding complex (CBC), which binds exclusively to capped RNA. The CBC is then recognized by the nuclear pore complex and exported. Once in the cytoplasm after the pioneer round of translation, the CBC complex is replaced by the translation factors eIF-4E and eIF-4G.

This complex is then recognized by other translation initiation machinery including the ribosome. Cap prevents 5' degradation in two ways. First, degradation of the mRNA by 5' exonucleases is prevented (as mentioned above) by functionally looking like a 3' end. Second, the CBC complex and the eIF-4E/eIF-4G block the access of decapping enzymes to the cap. This increases the half-life of the mRNA, essential in eukaryotes as the export process takes significant time.

Decapping of an mRNA is catalysed by the decapping complex made up of at least Dcp1 and Dcp2, which must compete with eIF-4E to bind the cap. Thus the 5' cap is a marker of an actively translating mRNA and is used by cells to regulate mRNA half-lives in response to new stimuli. Undesirable mRNAs are sent to P-bodies for temporary storage or decapping, the details of which are still being resolved.

The mechanism of 5' proximal intron excision promotion is not well understood, but the 5' cap appears to loop around and interact with thespliceosome in the splicing process, promoting intron excision.

Coding Regions

Coding regions are composed of codons, which are decoded and translated (in eukaryotes usually into one and in prokaryotes usually into several) proteins by the ribosome. Coding regions begin with the start codon and end with a stop codon. Generally, the start codon is an AUG triplet and the stop codon is UAA, UAG, or UGA.

The coding regions tend to be stabilised by internal base pairs, this impedes degradation. In addition to being protein-coding, portions of coding regions may serve as regulatory sequences in the pre-mRNA asexonic splicing enhancers or exonic splicing silencers.

Five Prime Untranslated Region

The five prime untranslated region (5' UTR), also known as the leader sequence, is a particular section of messenger RNA (mRNA) and the DNA that codes for it. At times the leader sequence can contain elements for control gene expression by way of attenuation.

The leader sequence will be translated and the availability of the encoded amino acids will affect the transcription of the rest of the mRNA. It starts at the +1 position (where transcription begins) and ends one nucleotide before the start codon (usually AUG) of the coding region. It usually contains a ribosome binding site (RBS), in bacteria also known as the Shine Dalgarno sequence (AGGAGGU). The 5' UTR may be a hundred or more nucleotides long, and the 3' UTR may be even longer (up to several kilobases in length).

An mRNA molecule codes for a protein through translation. The mRNA also contains regions that are not translated: in eukaryotes this includes the 5' untranslated region, 3' untranslated region, 5' cap and poly-A tail.

In prokaryotic mRNA the 5' UTR is normally short. Some viruses and cellular genes have unusual long structured 5' UTRs which may have roles in gene expression.

Several regulatory sequences may be found in the 5' UTR:

* Binding sites for proteins, that may affect the mRNA's stability or translation, for example iron responsive elements, which occur in the 5' UTRs (and 3' UTRs) of a small number ofeukaryotic mRNAs that regulate gene expression in response to iron.
* Regulatory elements that do not depend on proteins, such asriboswitches.
* Sequences that promote the initiation of translation.

Three Prime Untranslated Region

The three prime untranslated region (3' UTR) is a particular section of messenger RNA (mRNA). It follows the coding region.

An mRNA molecule codes for a protein through translation. The mRNA also contains regions that are not translated. In eukaryotes these regions are the cap, 5' untranslated region, 3' untranslated region, and polyA tail.

In prokaryotes mRNA structures have some differences (seemRNA) as do histone mRNAs. However, both have 3' UTRs.

Several regulatory sequences are found in the 3' UTR:

* A polyadenylation signal, usually *AAUAAA*, or a slight variant. This marks the site of cleavage of the transcript approximately 30 base pairs past the signal, followed by the addition of several hundred adenine residues (poly-A tail).
* Binding sites for proteins, that may affect the mRNA's stability or location in the cell, like SECIS elements (which direct the ribosome to translate the codon UGA as selenocysteines rather than as a stop codon), or AU-rich elements (AREs), stretches consisting of mainlyadenine and uracil nucleotides (which can either stabilize or destabilize the mRNA depending on the protein bound to it).
* Binding sites for miRNAs, a type of RNAi.

Polyadenylation

Polyadenylation is the addition of a poly(A) tail to an RNAmolecule. The poly(A) tail consists of multiple adenosine monophosphates; in other words, it is a stretch of RNA which only has Adenine bases. In eukaryotes, polyadenylation is part of the process that produces mature messenger RNA(mRNA) for translation. It therefore forms part of the larger process of gene expression.

The process of polyadenylation begins as the transcription of a gene finishes. The 3'-most segment of the newly-made RNA is first cleaved off by a set of proteins; these proteins then synthesise the poly(A) tail at the RNA's 3' end. In some genes these proteins may add a poly(A) tail at any one of several possible sites, polyadenylation can therefore produce more than one transcript from a single gene, similar to alternative splicing. The poly(A) tail is important for the nuclear export, translation and stability of mRNA. The tail is shortened over time and when it is short enough, the mRNA is enzymatically degraded.

However, in a few cell types, mRNAs with short poly(A) tails are stored for later activation by re-polyadenylation in the cytosol. In contrast, when polyadenylation occurs in bacteria, it promotes RNA degradation. This is also sometimes the case for eukaryotic non-coding RNAs. The wide distribution of polyadenylation among living organisms indicates that this process evolved early in the history of life on Earth.

Primer on RNA

RNAs are a type of large biological molecules, whose individual building blocks are called nucleotides. The name *poly(A) tail* (for polyadenylic acid tail) reflects the way RNA nucleotides are abbreviated, with a letter for the base the nucleotide contains (A for adenine, C for cytosine, G for guanine and U for uracil). RNAs are produced (*transcribed*) from a DNA template. By convention, RNA sequences are written in a 5' to 3' direction. The 5' end is the part of the RNA molecule that is transcribed first, and the 3' end is transcribed last. The 3' end is also where the poly(A) tail is found on polyadenylated RNAs.

Messenger RNA (mRNA) is RNA that has a coding region that acts as a template for protein synthesis (*translation*). The rest of the mRNA, the *untranslated regions*, tune how active the mRNA is. There are also many RNAs that are not translated, called non-coding RNAs. Like the untranslated regions, many of these non-coding RNAs have regulatory roles.

Nuclear Polyadenylation

Function

In nuclear polyadenylation, a poly(A) tail is added to an RNA at the end of transcription. On mRNAs, the poly(A) tail protects the mRNA molecule from enzymatic degradation in the cytoplasm and aids in transcription termination, export of the mRNA from the nucleus, and translation. Almost all eukaryotic mRNAs are polyadenylated, with the exception of animal replication-dependent histone mRNAs. These are the only mRNAs in eukaryotes that lack a poly(A) tail, ending instead in a stem-loop structure followed by a purine-rich sequence, termed histone downstream element, that directs where the RNA is cut so that the 3' end of the histone mRNA is formed.

Many eukaryotic non-coding RNAs are always polyadenylated at the end of transcription. There are small RNAs where the poly(A) tail is only seen in intermediary forms and not in the mature RNA as the ends are removed during processing, notably microRNAs. But for manylong noncoding RNAs – a seemingly large group of regulatory RNAs that for example includes the RNA Xist which mediates X chromosome inactivation – a poly(A) tail is part of the mature RNA.

Mechanism

The polyadenylation machinery in the nucleus of eukaryotes works on products of RNA polymerase II, such as precursor mRNA. Here, a multi-protein complex cleaves the 3'-most part of a newly produced RNA and polyadenylates the end produced by this cleavage. The cleavage is catalysed by the enzyme CPSF and occurs 10–30 nucleotides downstream of its binding site.This site is often the sequence AAUAAA on the RNA, but variants of it exist that bind more weakly to CPSF.

Two other proteins add specificity to the binding to an RNA: CstF and CFI. CstF binds to a GU-rich region further downstream of CPSF's site.CFI recognises a third site on the RNA (a set of UGUAA sequences in mammals) and can recruit CPSF even if the AAUAAA sequence is missing. The polyadenylation signal – the sequence motif recognised by the RNA cleavage complex – varies between groups of eukaryotes. Most human polyadenylation sites contain the AAUAAA sequence, but this sequence is less common in plants and fungi.

The RNA is cleaved right after transcription, as CstF also binds to RNA polymerase II. Cleavage also involves the protein CFII, unknown how. The cleavage site associated with a polyadenylation signal can vary up to some 50 nucleotides. When the RNA is cleaved, polyadenylation starts, catalysed by polyadenylate polymerase.

Polyadenylate polymerase builds the poly(A) tail by adding adenosine monophosphate units from adenosine triphosphate to the RNA, cleaving off pyrophosphate. Another protein, PAB2, binds to the new, short poly(A) tail and increases the affinity of polyadenylate polymerase for the RNA. When the poly(A) tail is approximately 250 nucleotides long the enzyme can no longer bind to CPSF and polyadenylation stops, thus determing the length of the poly(A) tail. CPSF is in contact with RNA polymerase II, allowing it to tell the polymerase to terminate transcription. The polyadenylation machinery is also physically linked to the spliceosome, a complex that removes introns from RNAs.

Downstream Effects

The poly(A) tail acts as the binding site for poly(A)-binding protein. Poly(A)-binding protein promotes export from the nucleus and translation, and inhibits degradation. This protein binds to the poly(A) tail prior to mRNA export from the nucleus and in yeast also recruits poly(A) nuclease, an enzyme that shortens the poly(A) tail and allows the export of the mRNA. Poly(A)-binding protein is exported to the cytoplasm with the RNA. mRNAs that are not exported are degraded by the exosome. Poly(A)-binding protein also can bind to, and thus recruit, several proteins that affect translation, one of these is initiation factor-4G which in turn recruits the 40Sribosomal subunit. However, a poly(A) tail is not required for the translation of all mRNAs.

Deadenylation

In eukaryotic somatic cells, the poly(A) tail of most mRNAs in the cytoplasm gradually get shorter, and mRNAs with shorter poly(A) tail are translated less and degraded sooner. However, it can take many hours before an mRNA is degraded. This deadenylation and degradation process can be accelerated by microRNAs complementary to the 3' untranslated region of an mRNA. In immature egg cells, mRNAs with shortened poly(A) tails are not degraded, but are instead stored without being translated. They are then activated by cytoplasmic polyadenylation after fertilisation, during egg activation.

In animals, poly(A) ribonuclease (PARN) can bind to the 5' cap and remove nucleotides from the poly(A) tail. The level of access to the 5' cap and poly(A) tail is important in controlling how soon the mRNA is degraded. PARN deadenylates less if the RNA is bound by the initiation factors 4E (at the 5' cap) and 4G (at the poly(A) tail), so this is why translation reduces deadenylation. The rate of deadenylation may also be regulated by RNA-binding proteins. Once the poly(A) tail is removed, the decapping complex removes the 5' cap, leading to a degradation of the RNA. Several other enzymes that seem to be involved in deadenylation have been identified in yeast.

Alternative Polyadenylation

Many protein-coding genes have more than one polyadenylation site, so a gene can code for several mRNAs that differ in their 3' end. Since alternative polyadenylation changes the length of the 3' untranslated region, it can change which binding sites for microRNAs the 3' untranslated region contains. MicroRNAs tend to repress translation and promote degradation of the mRNAs they bind to, although there are examples of microRNAs that stabilise transcripts. Alternative polyadenylation can also shorten the coding region, thus making the mRNA code for a different protein, but this is much less common than just shortening the 3' untranslated region.

The choice of poly(A) site depends on the expression of the proteins that take part in polyadenylation. For example, the expression of CstF-64, a subunit of cleavage stimulatory factor (CstF), increases in macrophages in response to lipopolysaccharides (a group of bacterial compounds that trigger an immune response). This results in the selection of weak poly(A) sites and thus shorter transcripts.

This removes regulatory elements in the 3' untranslated regions of mRNAs for defence-related products like lysozyme and TNF-α. These mRNAs then have longer half-lives and produce more of these proteins. RNA-binding proteins other than those in the polyadenylation machinery can also affect whether a polyadenyation site is used, as can DNA methylationnear the polyadenylation signal.

Cytoplasmic Polyadenylation

There is polyadenylation in the cytosol of some animal cell types, namely in the germ line, during early embryogenesis and in post-synapticsites of nerve cells. This lengthens the poly(A) tail of an mRNA with a shortened poly(A) tail, so that the mRNA will be translated.These shortened poly(A) tails are often less than 20 nucleotides, and are lengthened to around 80–150 nucleotides.

In the early mouse embryo, cytoplasmic polyadenylation of maternal RNAs from the egg cell allows the cell to survive and grow even though transcription does not start until the middle of the 2-cell stage (4-cell stage in human). In the brain, cytoplasmic polyadenylation is active during learning and could to play a role in long-term potentiation, which is the strengthening of the signal transmission from a nerve cell to another in response to nerve impulses and is important for learning and memory formation.

Cytoplasmic polyadenylation requires the RNA-binding proteins CPSF and CPEB, and can involve other RNA-binding proteins like Pumilio. Depending on the cell type, the polymerase can be the same type of polyadenylate polymerase (PAP) that is used in the nuclear process, or the cytoplasmic polymerase GLD-2.

Tagging for Degradation in Eukaryotes

For many non-coding RNAs, including tRNA, rRNA, snRNA and snoRNA, polyadenylation is a way of marking the RNA for degradation, in at least yeast. This polyadenylation is done in the nucleus by the TRAMP complex, which adds a tail that is around 40 nucleotides long to the 3' end. The RNA is then degraded by the exosome. Poly(A) tails have also been found on human rRNA fragments, both the form of homopolymeric (A only) and heterpolymeric (mostly A) tails.

In Prokaryotes and Organelles

In many bacteria, both mRNAs and non-coding RNAs can be polyadenylated. This poly(A) tail promotes degradation by the degradosome, which contains two RNA-degrading enzymes: polynucleotide phosphorylase and RNase E. Polynucleotide phosphorylase binds to the 3' end of RNAs and the 3' extension provided by the poly(A) tail allows it to bind to the RNAs whose secondary structure would otherwise block the 3' end. Successive rounds of polyadenylation and degradation of the 3' end by polynucleotide phosphorylase allows the degradosome to overcome these secondary structures. The poly(A) tail can also recruit RNases that cut the RNA in two. These bacterial poly(A) tails are about 30 nucleotides long.

In as different groups as animals and trypanosomes, the mitochondria contain both stabilising and destabilising poly(A) tails. Destabilising polyadenylation targets both mRNA and noncoding RNAs. The poly(A) tails are 43 nucleotides long on average. The stabilising ones start at the stop codon, and without them the stop codon (UAA) is not complete as the genome only encodes the U or UA part. Plant mitochondria only have destabilising polyadenylation, and yeast mitochondria have no polyadenylation at all.

While many bacteria and mitochondria have polyadenylate polymerases, they also have another type of polyadenylation, performed bypolynucleotide phosphorylase itself. This enzyme is found in bacteria, mitochondria, plastids and as a constituent of the archeal exosome (in those archaea that have an exosome). It can synthesise a 3' extension where the vast majority of the bases are adenines. Like in bacteria, polyadenylation by polynucleotide phosphorylase promotes degradation of the RNA in plastids and likely also archaea.

Evolution

Although polyadenylation is seen in almost all organisms, it is not universal. However, the wide distribution of this modification and the fact that it is present in organisms from all three domains of life implies that the last universal common ancestor of all living organisms probably had some form of polyadenylation system. A few organisms do not polyadenylate mRNA, which implies that they have lost their polyadenylation machineries during evolution. Although no examples of eukaryotes that lack polyadenylation are known, mRNAs from the bacterium *Mycoplasma gallisepticum* and the salt-tolerant archaean *Haloferax volcanii* lack this modification.

The most ancient polyadenylating enzyme is polynucleotide phosphorylase. This enzyme is part of both the bacterial degradosome and archaeal exosome, two closely related complexes that recycle RNA into nucleotides. This enzyme degrades RNA by attacking the bond between the 3'-most nucleotides with a phosphate, breaking off a diphosphate nucleotide.

This reaction is reversible, and so the enzyme can also extend RNA with more nucleotides. The heteropolymeric tail added by polynucleotide phosphorylase is very rich in adenine. The choice of adenine is most likely the result of higher ADP concentrations than other nucleotides as a result of using ATP as an energy currency, making it more likely to be incorporated in this tail in early lifeforms. It has been suggested that the involvement of adenine-rich tails in RNA degradation prompted the later evolution of polyadenylate polymerases (the enzymes that produce poly(A) tails with no other nucleotides in them).

Polyadenylate polymerases are not as ancient. They have separately evolved in both bacteria and eukaryotes from CCA-adding enzyme, which is the enzyme that completes the 3' ends of tRNAs. Its catalytic domain is homologous to that of other polymerases. Likely, the horizontal transfer of bacterial CCA-adding enzyme to eukaryotes allowed the archaeal-like CCA-adding enzyme to switch function to a poly(A) polymerase. Some lineages, like archaea and cyanobacteria, never evolved a polyadenylate polymerase.

History

Polyadenylation was first identified in 1960 as an enzymatic activity in extracts made from cell nuclei that could polymerise ATP, but not ADP, into polyadenine. Although identified in many types of cells, this activity had no known function until 1971, when poly(A) sequences were found in mRNAs. Initially, the only function of these sequences was thought to be protection of the 3' end of the RNA from nucleases, but later the specific roles of polyadenylation in nuclear export and translation were identified.

The polymerases responsible for polyadenylation were first purified and characterized in the 1960s and 1970s, but the large number of accessory proteins that control this process were only discovered in the early 1990s.

Monocistronic Versus Polycistronic mRNA

An mRNA molecule is said to be monocistronic when it contains the genetic information to translate only a single protein. This is the case for most of the eukaryotic mRNAs. On the other hand, polycistronic mRNA carries the information of several genes, which are translated into several proteins. These proteins usually have a related function and are grouped and regulated together in an operon. Most of the mRNA found in bacteria and archea are polycistronic. Dicistronic or bicistronic is the term used to describe an mRNA that encodes only twoproteins.

mRNA Circularization

In eukaryotes it is thought that mRNA molecules form circular structures due to an interaction between the cap binding complex and poly(A)-binding protein. Circularization is thought to promote recycling of ribosomes on the same message leading to efficient translation.

Degradation

Different mRNAs within the same cell have distinct lifetimes (stabilities). In bacterial cells, individual mRNAs can survive from seconds to more than an hour; in mammalian cells, mRNA lifetimes range from several minutes to days. The greater the stability of an mRNA, the more protein may be produced from that mRNA. The limited lifetime of mRNA enables a cell to alter protein synthesis rapidly in response to its changing needs. There are many mechanisms that lead to the destruction of a mRNA, some of which are described below.

Prokaryotic mRNA Degradation

In prokaryotes the lifetime of mRNA is generally much shorter than in eukaryotes. Prokaryotes degrade messages by using a combination of ribonucleases, including endonucleases, 3' exonucleases, and 5' exonucleases. In some instances, small RNA molecules (sRNA) tens to hundreds of nucleotides long can stimulate the degradation of specific mRNAs by base pairing with complementary sequences and facilitating ribonuclease cleavage. It was recently shown that bacteria also have a sort of 5' cap consisting of a triphosphate on the 5' end.Removal of two of the phosphates leaves a 5' monophosphate, causing the message to be destroyed by the endonuclease RNase E.

Eukaryotic mRNA Turnover

Inside eukaryotic cells there is a balance between the processes of translation and mRNA decay. Messages that are being actively translated are bound by ribosomes, the eukaryotic initiation factors eIF-4E and eIF-4G, and poly(A)-binding protein. eIF-4E and eIF-4G block the decapping enzyme (DCP2), and poly(A)-binding protein blocks the exosome complex, protecting the ends of the message. The balance between translation and decay is reflected in the size and abundance of cytoplasmic structures known as P-bodies.

The poly(A) tail of the mRNA is shortened by specialized exonucleases that are targeted to specific messenger RNAs by a combination of cis-regulatory sequences on the RNA and trans-acting RNA-binding proteins. Poly(A) tail removal is thought to disrupt the circular structure of the message and destabilize the cap binding complex. In this way, translationally inactive messages can be destroyed quickly, while active messages remain intact. The mechanism by which translation stops and the message is handed-off to decay complexes is not understood in detail.

AU-rich Element Decay

The presence of AU-rich elements in some mammalian mRNAs tends to destabilize those transcripts through the action of cellular proteins that bind these sequences and stimulate poly(A) tail removal. Loss of the poly(A) tail is thought to promote mRNA degradation by facilitating attack by both the exosome complex and the decapping complex. Rapid mRNA degradation via AU-rich elements is a critical mechanism for preventing the overproduction of potent cytokines such as tumour necrosis factor (TNF) and granulocyte-macrophage colony stimulating factor (GM-CSF). AU-rich elements also regulate the biosynthesis of proto-oncogenic transcription factors like c-Jun and c-Fos.

Nonsense Mediated Decay

Nonsense mediated decay (NMD) is a cellular mechanism of mRNA surveillance that functions to detect nonsense mutations and prevent the expression of truncated or erroneous proteins. Following transcription, precursor mRNA undergoes an assemblage of ribonucleoprotein(RNP) components followed by regulatory pre-mRNA processing.

Large average intron size in eukaryotic cells greatly increases the probability that aberrant mRNA splicing will result in the presence of a nonsense (stop) codon (UAA, UAG, UGA) somewhere within the open reading frame. NMD is triggered by exon junction complexes (EJCs) (components of the assembled RNP) that are deposited during pre-mRNA processing. While EJCs located in an open reading frame upstream of exon-exon junctions may facilitate ribosomal recruitmentprior to displacement by a "pioneer" round of translation, EJCs located downstream of a nonsense codon are not displaced because the ribosome is released from the transcript before reaching it.

These remaining EJCs function as tags for recruitment of UPF1 following the mRNA's transport out of the nucleus and into the cytosol where the RNA is degraded, for example by the exosome complex. NMD is not only a mechanism for degrading aberrant mRNA, however, as there are numerous examples of normal transcripts whose expression is regulated by this process including the plasticity protein Arc/Arg3.1.

Small Interfering RNA

Small interfering RNA (siRNA), sometimes known as short interfering RNA orsilencing RNA, is a class of double-stranded RNA molecules, 20-25 nucleotides in length, that play a variety of roles in biology. Most notably, siRNA is involved in theRNA interference (RNAi) pathway, where it interferes with the expression of a specific gene. In addition to their role in the RNAi pathway, siRNAs also act in RNAi-related pathways, e.g., as an antiviral mechanism or in shaping the chromatinstructure of a genome; the complexity of these pathways is only now being elucidated. siRNAs were first discovered by David Baulcombe's group at the Sainsbury Laboratory in Norwich, England, as part of post-transcriptional gene silencing(PTGS) in plants. The group published their findings in *Science* in a paper titled "A species of small antisense RNA in posttranscriptional gene silencing in plants".Shortly thereafter, in 2001, synthetic siRNAs were shown to be able to induce RNAi in mammalian cells by Thomas Tuschl, and colleagues in a paper published in*Nature*. This discovery led to a surge in interest in harnessing RNAi for biomedical research and drug development.

Structure

siRNAs have a well-defined structure: a short (usually 21-nt) double strand RNA (dsRNA) with 2-nt 3' overhangs on either end: Each strand has a 5' phosphate group and a 3' hydroxyl (-OH) group. This structure is the result of processing by dicer, an enzyme that converts either long dsRNAs or small hairpin RNAs into siRNAs. siRNAs can also be exogenously (artificially) introduced into cells by various transfection methods to bring about the specific knockdown of a gene of interest.

Essentially any gene of which the sequence is known can thus be targeted based on sequence complementarity with an appropriately tailored siRNA. This has made siRNAs an important tool for gene function and drug target validation studies in the post-genomic era.

RNAi Induction using SiRNAs or their Biosynthetic Precursors

Transfection of an exogenous siRNA can be problematic because the gene knockdown effect is only transient, particularly in rapidly dividing cells. One way of overcoming this challenge is to modify the siRNA in such a way as to allow it to be expressed by an appropriate vector, e.g., aplasmid. This is done by the introduction of a loop between the two strands, thus producing a single transcript, which can be processed into a functional siRNA. Such transcription cassettes typically use an RNA polymerase III promoter (e.g., U6 or H1), which usually directs the transcription of small nuclear RNAs (snRNAs) (U6 is involved in gene splicing; H1 is the RNasecomponent of human RNase P). It is assumed (although not known for certain) that the resulting siRNA transcript is then processed by Dicer.

RNA Activation

It has recently been found that dsRNA can also activate gene expression, a mechanism that has been termed "small RNA-induced gene activation" or RNAa. It has been shown that dsRNAs targeting gene promoters induce potent transcriptional activation of associated genes. RNAa was demonstrated in human cells using synthetic dsRNAs, termed "small activating RNAs" (saRNAs). It is currently not known whether RNAa is conserved in other organisms.

Challenges: Avoiding Nonspecific Effects

Because RNAi intersects with a number of other pathways, it is not surprising that on occasion nonspecific effects are triggered by the experimental introduction of an siRNA. When a mammalian cell encounters a double-stranded RNA such as an siRNA, it may mistake it as a viral by-product and mount an immune response. Furthermore, because structurally related microRNAs modulate gene expression largely via incomplete complementarity base pair interactions with a target mRNA, the introduction of an siRNA may cause unintended off-targeting.

Innate Immunity

Introduction of too much siRNA can result in nonspecific events due to activation of innate immune responses. Most evidence to date suggests that this is probably due to activation of the dsRNA sensor PKR, although retinoic acid inducible gene I (RIG-I) may also be involved. The induction of cytokines via toll-like receptor 7 (TLR7) has also been described. One promising method of reducing the nonspecific effects is to convert the siRNA into a microRNA. MicroRNAs occur naturally, and by harnessing this endogenous pathway it should be possible to achieve similar gene knockdown at comparatively low concentrations of resulting siRNAs. This should minimize nonspecific effects.

Off-targeting

Off-targeting is another challenge to the use of siRNAs as a gene knockdown tool. Here, genes with incomplete complementarity are inadvertently downregulated by the siRNA (effectively, the siRNA acts as a miRNA), leading to problems in data interpretation and potential toxicity.

This, however, can be partly addressed by designing appropriate control experiments, and siRNA design algorithms are currently being developed to produce siRNAs free from off-targeting. Genome-wide expression analysis, e.g., by microarray technology, can then be used to verify this and further refine the algorithms. A 2006 paper from the laboratory of Dr. Khvorova implicates 6- or 7-basepair-long stretches from position 2 onward in the siRNA matching with 3'UTR regions in off-targeted genes.

Possible Therapeutic Applications and Challenges

Given the ability to knock down essentially any gene of interest, RNAi via siRNAs has generated a great deal of interest in both basic and applied biology. There are an increasing number of large-scale RNAi screens that are designed to identify the important genes in various biological pathways. Because disease processes also depend on the activity of multiple genes, it is expected that in some situations turning off the activity of a gene with an siRNA could produce a therapeutic benefit.

However, applying RNAi via siRNAs to living animals, especially humans, poses many challenges. Experimentally, siRNAs show different effectiveness in different cell types in a manner as yet poorly understood: some cells respond well to siRNAs and show a robust knockdown, whereas others show no such knockdown (even despite efficient transfection). Phase I results of the first two therapeutic RNAi trials (indicated for age-related macular degeneration, aka AMD) reported at the end of 2005 that siRNAs are well tolerated and have suitable pharmacokinetic properties.

MicroRNA

MicroRNAs (miRNAs) are short ribonucleic acid (RNA) molecules, on average only 22 nucleotides long and are found in all eukaryotic cells. miRNAs are post-transcriptional regulators that bind to complementary sequences on target messenger RNA transcripts (mRNAs), usually resulting in translational repression and gene silencing. The human genome may encode over 1000 miRNAs, which may target about 60% of mammalian genes and are abundant in many human cell types.

miRNAs show very different characteristics between plants and metazoans. In plants the miRNA complementarity to its mRNA target is nearly perfect, with no or few mismatched bases. In metazoans on the other hand miRNA complementarity is far from perfect and one miRNA can target many different sites on the same mRNA or on many different mRNAs. Another difference is the location of target sites on mRNAs. In metazoans the miRNA target sites are in the three prime untranslated regions (3'UTR) of the mRNA. In plants targets can be located in the 3' UTR but are more often in the coding region itself.MiRNAs are well conserved in eukaryotic organism and are thought to be a vital and evolutionary ancient component of genetic regulation.

The first miRNAs were characterized in the early 1990s, but miRNAs were not recognized as a distinct class of biologic regulators with conserved functions until the early 2000s. Since then, miRNA research has revealed multiple roles in negative regulation (transcript degradation and sequestering, translational suppression) and possible involvement in positive regulation (transcriptional and translational activation). By affecting gene regulation, miRNAs are likely to be involved in most biologic processes. Different sets of expressed miRNAs are found in different cell types and tissues. Aberrant expression of miRNAs has been implicated in numerous disease states, and miRNA-based therapies are under investigation.

History

MicroRNAs were discovered in 1993 by Victor Ambros, Rosalind Lee and Rhonda Feinbaum during a study of the gene *lin-14* in *C. elegans*development.

They found that LIN-14 protein abundance was regulated by a short RNA product encoded by the *lin-4* gene. A 61 nucleotide precursor from *lin-4* gene matured to a 22 nucleotide RNA containing sequences partially complementary to multiple sequences in the 3' UTR of the *lin-14* mRNA. This complementarity was sufficient and necessary to inhibit the translation of *lin-14* mRNA into LIN-14 protein. Retrospectively, the *lin-4* small RNA was the first microRNA to be identified, though at the time, it was thought to be a nematode idiosyncrasy. Only in 2000 was a second RNA characterized: let-7, which repressed *lin-41*, *lin-14*, *lin-28*, *lin-42*, and *daf-12* expression during developmental stage transitions in *C. elegans*. let-7 was soon found to be conserved in many species, indicating the existence of a wider phenomenon.

Nomenclature

Under a standard nomenclature system, names are assigned to experimentally confirmed miRNAs before publication of their discovery.The prefix "mir" is followed by a dash and a number, the latter often indicating order of naming. For example, mir-123 was named and likely discovered prior to mir-456.

The uncapitalized "mir-" refers to the pre-miRNA, while a capitalized "miR-" refers to the mature form. miRNAs with nearly identical sequences bar one or two nucleotides are annotated with an additional lower case letter. For example, miR-123a would be closely related to miR-123b. Pre-miRNAs that lead to 100% identical mature miRNAs but that are located at different places in the genome are indicated with an additional dash-number suffix. For example, the pre-miRNAs hsa-mir-194-1 and hsa-mir-194-2 lead to an identical mature miRNA (hsa-miR-194) but are located in different regions of the genome. Species of origin is designated with a three-letter prefix, e.g., hsa-miR-123 would be from human (*Homo sapiens*) and oar-miR-123 would be a sheep (*Ovis aries*) miRNA.

Other common prefixes include 'v' for viral (miRNA encoded by a viral genome) and 'd' for *Drosophila* miRNA (a fruit fly commonly studied in genetic research). When two mature microRNAs originate from opposite arms of the same pre-miRNA, they are denoted with a -3p or -5p suffix. (In the past, this distinction was also made with 's' (sense) and 'as' (antisense)). When relative expression levels are known, an asterisk following the name indicates an miRNA expressed at low levels relative to the miRNA in the opposite arm of a hairpin. For example, miR-123 and miR-123* would share a pre-miRNA hairpin, but more miR-123 would be found in the cell.

Biogenesis

Most microRNA genes are found in intergenic regions or in anti-sense orientation to genes and contain their own miRNA gene promoter and regulatory units. As much as 40% of miRNA genes may lie in the introns of protein and non-protein coding genes or even in exons. These are usually, though not exclusively, found in a sense orientation and thus usually are regulated together with their host genes.

Other miRNA genes showing a common promoter include the 42-48% of all miRNAs originating from polycistronic units containing 2-7 discrete loops from which mature miRNAs are processed, although this does not necessarily mean the mature miRNAs of a family will be homologous in structure and function. The promoters mentioned have been shown to have some similarities in their motifs to promoters of other genes transcribed by RNA polymerase II such as protein coding genes. The DNA template is not the final word on mature miRNA production: 6% of human miRNAs show RNA editing, the site-specific modification of RNA sequences to yield products different from those encoded by their DNA. This increases the diversity and scope of miRNA action beyond that implicated from the genome alone.

Transcription

miRNA genes are usually transcribed by RNA polymerase II (Pol II). The polymerase often binds to a promoter found near the DNA sequence encoding what will become the hairpin loop of the pre-miRNA. The resulting transcript is capped with a specially-modified nucleotide at the 5' end, polyadenylated with multiple adenosines (a poly(A) tail), and spliced.

The product, called a primary miRNA (pri-miRNA), may be hundreds or thousands of nucleotides in length and contain one or more miRNA stem loops. When a stem loop precursor is found in the 3' UTR, a transcript may serve as a pri-miRNA and a mRNA. RNA polymerase III (Pol III) transcribes some miRNAs, especially those with upstream Alu sequences, transfer RNAs (tRNAs), and mammalian wide interspersed repeat (MWIR) promoter units.

Nuclear Processing

A single pri-miRNA may contain from one to six miRNA precursors. These hairpin loop structures are composed of about 70 nucleotides each. Each hairpin is flanked by sequences necessary for efficient processing.

The double-stranded RNA structure of the hairpins in a pri-miRNA is recognized by a nuclear protein known as DiGeorge Syndrome Critical Region 8 (DGCR8 or "Pasha" in invertebrates), named for its association with DiGeorge Syndrome. DGCR8 associates with the enzyme Drosha, a protein that cuts RNA, to form the "Microprocessor" complex. In this complex, DGCR8 orients the catalytic RNase III domain of Drosha to liberate hairpins from pri-miRNAs by cleaving RNA about eleven nucleotides from the hairpin base (two helical RNA turns into the stem). The resulting hairpin, known as a pre-miRNA (precursor-miRNA), has a two-nucleotide overhang at its 3' end; it has 3' hydroxyl and 5' phosphate groups. Pre-miRNAs that are spliced directly out of introns, bypassing the Microprocessor complex, are known as "mirtrons." Originally thought to exist only in *Drosophila* and *C. elegans*, mirtrons have now been found in mammals.

Perhaps as many as 16% of pri-miRNAs may be altered through nuclear RNA editing. Most commonly, enzymes known as adenosine deaminases acting on RNA (ADARs) catalyse adenosine to inosine (A to I) transitions. RNA editing can halt nuclear processing (for example, of pri-miR-142, leading to degradation by the ribonuclease Tudor-SN) and alter downstream processes including cytoplasmic miRNA processing and target specificity (e.g., by changing the seed region of miR-376 in the central nervous system).

Nuclear Export

pre-miRNA hairpins are exported from the nucleus in a process involving the nucleocytoplasmic shuttle Exportin-5. This protein, a member of the *karyopherin* family, recognizes a two-nucleotide overhang left by the RNase III enzyme Drosha at the 3' end of the pre-miRNA hairpin. Exportin-5-mediated transport to the cytoplasm is energy-dependent, using GTP bound to the Ran protein.

Cytoplasmic Processing

In the cytoplasm, the pre-miRNA hairpin is cleaved by the RNase III enzyme Dicer. This endoribonuclease interacts with the 3' end of the hairpin and cuts away the loop joining the 3' and 5' arms, yielding an imperfect miRNA:miRNA* duplex about 22 nucleotides in length.

Overall hairpin length and loop size influence the efficiency of Dicer processing, and the imperfect nature of the miRNA:miRNA* pairing also affects cleavage. Although either strand of the duplex may potentially act as a functional miRNA, only one strand is usually incorporated into the RNA-induced silencing complex (RISC) where the miRNA and its mRNA target interact.

Biogenesis in Plants

miRNA biogenesis in plants differs from metazoan biogenesis mainly in the steps of nuclear processing and export. Instead of being cleaved by two different enzymes, once inside and once outside the nucleus, both cleavages of the plant miRNA is performed by a Dicer homolog, called Dicer-like1 (DL1).

DL1 is only expressed in the nucleus of plant cells, which indicates that both reactions take place inside the nucleus. Before plant miRNA:miRNA* duplexes are transported out of the nucleus its 3' overhangs are methylated by a RNA methyltransferaseprotein called Hua-Enhancer1 (HEN1). The duplex is then transported out of the nucleus to the cytoplasm by a protein called

Hasty (HST), an Exportin 5 homolog, where they disassemble and the mature miRNA is incorporated into the RISC.

RNA-induced Silencing Complex

RNA-Induced Silencing Complex, or RISC, is a multiprotein complex that incorporates one strand of a small interfering RNA (siRNA) ormicro RNA (miRNA). RISC uses the siRNA or miRNA as a template for recognizing complementary mRNA. When it finds a complementary strand, it activates RNase and cleaves the RNA. This process is important both in gene regulation by microRNAs and in defence against viral infections, which often use double-stranded RNA as an infectious vector.

Strands

The RNA endonuclease Dicer plays a role in aiding RISC action by providing the initial RNA material to activate the complex as well as the first RNA substrate molecule. When Dicer, which has endonuclease activity against dsRNA and pre-miRNAs, cleaves a pre-miRNA stem-loop or a dsRNA, a 20- to 25-base-pair double-stranded RNA fragment is formed with a two-nucleotide 3' overhang at each end.

- One strand is integrated into the RISC complex. This strand is known as the guide strand and is selected by the argonaute protein, the catalytically active RNase in the RISC complex, on the basis of the thermodynamic stability of the 5' end. The strand with a less thermodynamically stable 5' end is loaded into the RISC complex.
- The remaining strand, known as the anti-guide or passenger strand, is degraded as a RISC complex substrate.

Binding

The RISC complex with a bound siRNA recognizes complementary messenger RNA (mRNA) molecules and degrades them, resulting in substantially decreased levels of protein translation and effectively turning off the gene. It is as yet unclear how the activated RISC complex locates the mRNA targets in the cell, though it has been shown that the process can occur in situations outside of ongoing protein translation from mRNA.

Endogenously expressed miRNA is usually imperfectly complementary to a large number of nuclear genes and has a modulating effect on these genes' levels of expression via translational repression.

Mode of Silencing

Gene silencing may occur either via mRNA degradation or preventing mRNA from being translated. It has been demonstrated that if there is complete complementation between the miRNA and target mRNA sequence, Ago2 can cleave the mRNA and lead to direct mRNA degradation. Yet, if there isn't complete complementation the silencing is achieved by preventing translation.

miRNA Turnover

Turnover of mature miRNA is needed for rapid changes in miRNA expression profiles. During miRNA maturation in the cytoplasm, uptake by the Argonaute protein is thought to stabilize the guide strand, while the opposite (* or "passenger") strand is preferentially destroyed. In what has been called a "Use it or lose it" strategy, Argonaute may preferentially retain miRNAs with many targets over miRNAs with few or no targets, leading to degradation of the non-targeting molecules.

Decay of mature miRNAs in animals is mediated by the 5′-to-3′ exoribonuclease XRN2, also known as Rat1p. In plants, SDN (small RNA degrading nuclease) family members degrade miRNAs in the opposite (3'-to-5') direction. Similar enzymes are encoded in animal genomes, but their roles have not yet been described. Several miRNA modifications affect miRNA stability. As indicated by work in the model organism *Arabidopsis thaliana* (thale cress), mature plant miRNAs appear to be stabilized by the addition of methyl moieties at the 3' end. The 2'-O-conjugated methyl groups block the addition of uracil (U) residues by uridyltransferase enzymes, a modification that may be associated with miRNA degradation. However, uridylation may also protect some miRNAs; the consequences of this modification are incompletely understood.

Uridylation of some animal miRNAs has also been reported. Both plant and animal miRNAs may be altered by addition of adenine (A) residues to the 3' end of the miRNA. An extra A added to the end of mammalian miR-122, a liver-enriched miRNA important in Hepatitis C, stabilizes the molecule, and plant miRNAs ending with an adenine residue have slower decay rates.

Cellular Functions

The function of miRNAs appears to be in gene regulation. For that purpose, a miRNA is complementary to a part of one or more messenger RNAs (mRNAs). Animal miRNAs are usually complementary to a site in the 3' UTR whereas plant miRNAs are usually complementary to coding regions of mRNAs. Perfect or near perfect base pairing with the target RNA promotes cleavage of the RNA. This is the primary mode of plant microRNAs.

In animals, microRNAs more often only partially base pair and inhibit protein translation of the target mRNA(this exists in plants as well but is less common). MicroRNAs that are partially complementary to the target can also speed updeadenylation, causing mRNAs to be degraded sooner. For partially complementary microRNA to recognise their targets, the nucleotides 2–7 of the miRNA ('seed region') still have to be perfectly complementary. miRNAs occasionally also causes histone modification andDNA methylation of promoter sites and therefore affecting the expression of targeted genes.

Animal microRNAs target in particular developmental genes. In contrast, genes involved in functions common to all cells, such as gene expression, have very few microRNA target sites and seem to be under selection to avoid targeting by microRNAs. dsRNA can also activate gene expression, a mechanism that has been termed "small RNA-induced gene activation" or RNAa. dsRNAs targeting gene promoters can induce potent transcriptional activation of associated genes. This was demonstrated in human cells using synthetic dsRNAs termed small activating RNAs (saRNAs), but has also been demonstrated for endogenous microRNA.

Evolution

MicroRNAs are significant phylogenetic markers because of their astonishingly low rate of evolution. Their origin may have permitted the development of morphological innovation, and by making gene expression more specific and 'fine-tunable', permitted the genesis of complex organs and perhaps, ultimately, complex life. Indeed, rapid bursts of morphological innovation are generally associated with a high rate of microRNA accumulation. MicroRNAs originate predominantly by the random formation of hairpins in "non-coding" sections of DNA (i.e. introns or intergene regions), but also by the duplication and modification of existing microRNAs.

The rate of evolution (i.e. nucleotide substitution) in recently-originated microRNAs is comparable to that elsewhere in the non-coding DNA, implying evolution by neutral drift; however, older microRNAs have a much lower rate of change (often less than one substitution per hundred million years), suggesting that once a microRNA gains a function it undergoes extreme purifying selection.

At this point, a microRNA is rarely lost from an animal's genome, although microRNAs which are more recently derived (and thus presumably non-functional) are frequently lost. This makes them a valuable phylogenetic marker, and they are being looked upon as a possible solution to such outstanding phylogenetic problems as the relationships of arthropods. MicroRNAs feature in the genomes of most eukaryotic organisms, from the brown algae to the metazoa. Across all species, in excess of 5000 had been identified by March 2010. Whilst short RNA sequences (50 – hundreds of base pairs) of a broadly comparable function occur in bacteria, bacteria lack true microRNAs.

Experimental Detection and Manipulation of miRNA

MicroRNA expression can be quantified in a two-step polymerase chain reaction process of modified RT-PCR followed by quantitative real-time PCR.

Variations of this method achieve absolute or relative quantification. miRNAs can also be hybridized to microarrays, slides or chips with probes to hundreds or thousands of miRNA targets, so that relative levels of miRNAs can be determined in different samples. The activity of an miRNA can be experimentally inhibited using a locked nucleic acid (LNA) oligo, a Morpholino oligo or a 2'-O-methyl RNA oligo. MicroRNA maturation can be inhibited at several points by steric-blocking oligos. The miRNA target site of an mRNA transcript can also be blocked by a steric-blocking oligo. Additionally, a specific miRNA can be silenced by a complementaryantagomir. For the "in situ" detection of miRNA, the use of LNA is currently the only efficient method. The locked conformation of LNA results in enhanced hybridization properties and increases sensitivity and selectivity, making it ideal for detection of short miRNA.

miRNA and Disease

Just as miRNA is involved in the normal functioning of eukaryotic cells, so has dysregulation of miRNA been associated with disease. A manually curated, publicly available database miR2Disease documents known relationships between miRNA dysregulation and human disease.

miRNA and Cancer

Several miRNAs have been found to have links with some types of cancer.

A study of mice altered to produce excess c-Myc — a protein with mutated forms implicated in several cancers — shows that miRNA has an effect on the development of cancer. Mice that were engineered to produce a surplus of types of miRNA found in lymphoma cells developed the disease within 50 days and died two weeks later. In contrast, mice without the surplus miRNA lived over 100 days. Leukemia can be caused by the insertion of a viral genome next to the 17-92 array of microRNAs leading to increased expression of this microRNA.

Another study found that two types of miRNA inhibit the E2F1 protein, which regulates cell proliferation. miRNA appears to bind to messenger RNA before it can be translated to proteins that switch genes on and off.

By measuring activity among 217 genes encoding miRNA, patterns of gene activity that can distinguish types of cancers can be discerned. miRNA signatures may enable classification of cancer. This will allow doctors to determine the original tissue type which spawned a cancer and to be able to target a treatment course based on the original tissue type. miRNA profiling has already been able to determine whether patients with chronic lymphocytic leukemia had slow growing or aggressive forms of the cancer.

Transgenic mice that over-express or lack specific miRNAs have provided insight into the role of small RNAs in various malignancies. A novel miRNA-profiling based screening assay for the detection of early-stage colorectal cancer has been developed and is currently in clinical trials. Early results showed that blood plasma samples collected from patients with early, resectable (Stage II) colorectal cancer could be distinguished from those of sex-and age-matched healthy volunteers. Sufficient selectivity and specificity could be achieved using small (less than 1 mL) samples of blood. The test has potential to be a cost-effective, non-invasive way to identify at-risk patients who should undergo colonoscopy.

miRNA and Heart Disease

The global role of miRNA function in the heart has been addressed by conditionally inhibiting miRNA maturation in the murine heart, and has revealed that miRNAs play an essential role during its development. miRNA expression profiling studies demonstrate that expression levels of specific miRNAs change in diseased human hearts, pointing to their involvement in cardiomyopathies. Furthermore, studies on specific miRNAs in animal models have identified distinct roles for miRNAs both during heart development and under pathological conditions, including the regulation of key factors important for cardiogenesis, the hypertrophic growth response, and cardiac conductance.

miRNA and the Nervous System

miRNAs appear to regulate the nervous system. Neural miRNAs are involved at various stages of synaptic development, including dendritogenesis (involving miR-132, miR-134 and miR-124), synapse formation and synapse maturation (where miR-134 and miR-138 are thought to be involved). Some studies find altered miRNA expression in schizophrenia.

miRNA and Non-coding RNAs

When the human genome project mapped its first chromosome in 1999, it was predicted the genome would contain over 100,000 protein coding genes. However, only around 20,000 were eventually identified (International Human Genome Sequencing Consortium, 2004).

Since then, the advent of bioinformatics approaches combined with genome tiling studies examining the transcriptome, systematic sequencing of full length cDNA libraries, and experimental validation (including the creation of miRNA derived antisense oligonucleotides called antagomirs) have revealed that many transcripts are non protein-coding RNA, including several snoRNAs and miRNAs.

4

Ribosomal RNA

Ribosomal ribonucleic acid (rRNA) is the RNA component of the ribosome, the protein manufacturing organelle of all living cells. Ribosomal RNA provides a mechanism for decoding mRNA into amino acids and interacts with tRNAs during translation by providing peptidyl transferase activity. The tRNAs bring the necessary amino acids corresponding to the appropriate mRNA codon.

Inside the Ribosome

The ribosomal RNAs form two subunits, the large subunit (LSU) and small subunit (SSU). mRNA is sandwiched between the small and large subunits and the ribosome catalyses the formation of a peptide bond between the 2 amino acids that are contained in the rRNA.

A ribosome also has 3 binding sites called A, P, and E.

- The A site in the ribosome binds to an aminoacyl-tRNA (a tRNA bound to an amino acid).
- The amino (NH2) group of the aminoacyl-tRNA, which contains the new amino acid, attacks the ester linkage of peptidyl-tRNA (contained within the P site), which contains the last amino acid of the growing chain, forming a new peptide bond. This reaction is catalysed bypeptidyl transferase.
- The tRNA that was holding on the last amino acid is moved to the E site, and what used to be the aminoacyl-tRNA is the peptidyl-tRNA.

A single mRNA can be translated simultaneously by multiple ribosomes.

Prokaryotes

In prokaryotes a small 30S ribosomal subunit contains the 16S rRNA. The large 50S ribosomal subunit contains two rRNA species (the 5S and 23S rRNAs). Bacterial 16S, 23S, and 5S rRNA genes are typically organized as a co-transcribed operon.

There may be one or more copies of the operon dispersed in the genome (for example, *Escherichia coli* has seven). Archaea contains either a single rDNA operon or multiple copies of the operon.

The 3' end of the 16S rRNA (in a ribosome) binds to a sequence on the 5' end of mRNA called the Shine-Dalgarno sequence.

Eukaryotes

In contrast, eukaryotes generally have many copies of the rRNA genes organized in tandem repeats; in humans approximately 300–400 rDNA repeats are present in five clusters (on chromosomes 13, 14, 15, 21 and 22).

The 18S rRNA in most eukaryotes is in the small ribosomal subunit, and the large subunit contains three rRNA species (the 5S, 5.8S and 28S rRNAs). Mammalian cells have 2 mitochondrial (12S and 16S) rRNA molecules and 4 types of cytoplasmic rRNA (28S, 5.8S, 5S (large ribosome subunit) and 18S (small subunit)). 28S, 5.8S, and 18S rRNAs are encoded by a single transcription unit (45S) separated by 2 internally transcribed spacers.

The 45S rDNA organized into 5 clusters (each has 30-40 repeats) on chromosomes 13, 14, 15, 21, and 22. These are transcribed by RNA polymerase I. 5S occurs in tandem arrays (~200-300 true 5S genes and many dispersed pseudogenes), the largest one on the chromosome 1q41-42. 5S rRNA is transcribed by RNA polymerase III.

The tertiary structure of the small subunit ribosomal RNA (SSU rRNA) has been resolved by X-ray crystallography. The secondary structure of SSU rRNA contains 4 distinct domains — the 5', central, 3' major and 3' minor domains. A model of the secondary structure for the 5' domain (500-800 nucleotides) is shown.

Prokaryotes vs. Eukaryotes

Both prokaryotic and eukaryotic ribosomes can be broken down into two subunits (the S in 16S represents Svedberg units):

Type	*Size*	*Large subunit*	*Small subunit*
prokaryotic	70S	50S (5S, 23S)	30S (16S)
eukaryotic	80S	60S (5S, 5.8S, 28S)	40S (18S)

Note that the S units of the subunits cannot simply be added because they represent measures of sedimentation rate rather than of mass. The sedimentation rate of each subunit is affected by its shape, as well as by its mass.

Translation

Translation is the net effect of proteins being synthesized by ribosomes, from a copy (mRNA) of the DNA template in the nucleus. One of the components of the ribosome (16S rRNA) base pairs complementary to a sequence upstream of the start codon in mRNA.

Importance of rRNA

Ribosomal RNA characteristics are important in medicine and in evolution.

- rRNA is the target of several clinically relevant antibiotics: chloramphenicol, erythromycin, kasugamycin, micrococcin, paromomycin,ricin, sarcin, spectinomycin, streptomycin, and thiostrepton.
- rRNA is the one of the only genes present in all cells. For this reason, genes that encode the rRNA (rDNA) are sequenced to identify an organism's taxonomic group, calculate related groups, and estimate rates of species divergence. For this reason many thousands of rRNA sequences are known and stored in specialized databases such as RDP-II and SILVA.

Signal Recognition Particle RNA

The signal recognition particle RNA, also known as 7SL, 6S, or 4.5S RNA, is the RNA component of the signal recognition particle (SRP) ribonucleoprotein complex. SRP is a universally conserved ribonucleoprotein that directs the traffic of proteins within the cell and allows them to be secreted. The SRP RNA, together with one or more SRP proteins contributes to the binding and release of the signal peptide. The RNA and protein components of this complex are highly conserved but do vary between the different kingdoms of life.

The common SINE family Alu probably originated from a duplication of a 7SL RNA gene.

The eukaryotic SRP consists of a 300-nucleotide 7S RNA and six proteins: SRPs 72, 68, 54, 19, 14, and 9. Archaeal SRP consists of a 7S RNA and homologues of the eukaryotic SRP19 and SRP54 proteins. Eukaryotic and archaeal 7S RNAs have very similar secondary structures.

In most bacteria, the SRP consists of an RNA molecule (4.5S) and the Ffh protein (a homologue of the eukaryotic SRP54 protein). Some Gram-positive bacteria (e.g. *Bacillus subtilis*) have a longer eukaryote-like SRP RNA that includes an Alu domain. In eukaryotes and archaea, eight helical elements fold into theAlu and S domains, separated by a long linker region. The Alu domain is thought to mediate the peptide chain elongation retardation function of the SRP.

The universally conserved helix which interacts with the SRP54 M domain mediates signal sequence recognition. The SRP19-helix 6 complex is thought to be involved in SRP assembly and stabilises helix 8 for SRP54. binding The human genome in particular is known to contain a large amount of SRP RNA related sequence, including Alu repeats.

Discovery

SRP RNA was first detected in avian and murine oncogenicRNA (ocorna) virus particles. Subsequently, SRP RNA was found to be a stable component of uninfected HeLa cells where it associated with membrane and polysome fractions. In 1980, cell biologists purified from canine pancreas an 11S "signal recognition protein" (fortuitously also abbreviated "SRP") which promoted the translocation of secretory proteins across the membrane of the endoplasmic reticulum. It was then discovered that SRP contained an RNA component. Comparing the SRP RNA genes from different species revealed helix 8 of the SRP RNA to be highly conserved in all domains of life.

The regions near the 5'- and 3'-ends of the mammalian SRP RNA are similar to the dominant Alu family of middle repetitive sequences of the human genome. It is now understood that Alu DNA originated from SRP RNA by excision of the central SRP RNA-specific (S) fragment, followed by reverse transcription and integration into multiple sites of the human chromosomes. SRP RNAs have been identified also in some organelles, for example in the plastid SRPs of many photosynthetic organisms.

Transcription and Processing

Eukaryotic SRP RNAs are transcribed from DNA by RNA polymerase III. RNA polymerase III also transcribes the genes for 5Sribosomal RNA, tRNA, 7SK RNA, and U6 spliceosomal RNA.

The promoters of the human SRP RNA genes include elements located downstream of the transcriptional start site. Plant SRP RNA promoters contain an upstream stimulatory element (USE) and a TATA box. Yeast SRP RNA genes have a TATA box and intragenic sequences (A- and B-blocks) which likely play a role intranscription.

In the bacteria, genes are organized in operons and transcribed by RNA polymerase. The 5'-end of the small (4.5S) SRP RNA of many bacteria is cleaved by RNase P. The ends of the Bacillus subtilis SRP RNA are processed by RNase III. So far, no SRP RNA introns have been observed.

Function

Co-translational Translocation

The SRP RNA is an integral part of the small and the large domain of the SRP. The function of the small domain is to delay protein translation until the ribosome-bound SRP has an opportunity to associate with the membrane-resident SRP receptor (SR). Within the large domain, the SRP RNA of the signal peptide-charged SRP promotes the hydrolysis of two guanosine triphosphate (GTP) molecules. This reaction releases the SRP from the SRP receptor and the ribosome, allowing translation to continue and the proteinto enter the translocon. The protein transverses the membrane co-translationally (during translation) and enters into another cellular compartment or the extracellular space. In eukaryotes, the target is the membrane of the endoplasmic reticulum (ER). InArchaea, SRP delivers proteins to the plasma membrane. In thebacteria, SRP primarily incorporates proteins into the inner membrane.

Post-translational Transport

SRP participates also in the sorting of proteins after their synthesis has been completed (post-translational protein sorting). Ineukaryotes, tail-anchored proteins possessing a hydrophobic insertion sequence at their C-terminus are delivered to the endoplasmic reticulum (ER) by the SRP. Similarly, the SRP assists post-translationally in the import of nuclear encoded proteins to the thylakoidmembrane of chloroplasts.

Structure

In 2005, a nomenclature for all SRP RNAs proposed a numbering system of 12 helices. Helix sections are named with a lower case letter suffix (e.g. 5a). Insertions, or helix “branches” are given dotted numbers (e.g. 9.1 and 12.1).

The SRP RNA spans a wide phylogenetic spectrum with respect to size and the number of its structural features. The smallest functional SRP RNAs have been found in mycoplasma and related species. *Escherichia coli* SRP RNA (also called 4.5S RNA) is composed of 114 nucleotide residues and forms an RNA stem-loop. The gram-positive bacterium *Bacillus subtilis* encodes a larger 6S SRP RNA which resemble the Archaeal homologs but lacks SRP RNA helix 6.

Archaeal SRP RNAs possess helices 1 to 8, lack helix 7, and are characterized by a tertiary structure which involves the apicalloops of helix 3 and helix 4. The eukaryotic SRP RNAs lack helix 1 and contain a helix 7 of variable size. Some protozoan SRP RNAs have reduced helices 3 and 4. The ascomycota SRP RNAs have an all together reduced small domain and lack helices 3 and 4. The largest SRP RNAs known to date are found in the yeasts (Saccharomycetes) which acquired helices 9 to 12 as insertions into helix 5, as well as an extended helix 7. Seed plants express numerous highly divergent SRP RNAs.

Motifs

Four conserved features (motifs) have been identified (shown in the Figure in dark gray): the (1) SRP54 binding motif, (2) Helix 6 GNAR tetraloop motif, (3) 5e motif, and (4) UGU(NR) motif.

SRP54 Binding

The asymmetric loop between helical sections 8a and 8b and the adjacent base paired 8b section are a prominent property of every SRP RNA. Helical section 8b contains non-Watson-Crick base pairings which contribute to the formation of a flatted minor groove in the RNA suitable for the binding of protein SRP54 (called Ffh in the bacteria). The apical loop of helix 8 contains four, five, or six residues, depending on the species. It has a highly conserved guanosine as the first and an adenosine as the last loop residue. This feature is required for the interaction with the third adenosine residue of the helix 6 GNAR tetraloop motif.

Helix 6 GNAR Tetraloop

The SRP RNAs of eukaryotes and Archaea have a GNAR tetraloop (N is for any nucleotide, R is for a purine) in helix 6. Its conservedadenosine residue is important for the binding of protein SRP19. This adenosine makes a tertiary interaction with another adenosineresidue located in the apical loop of helix 8.

5e

The 11 nucleotides of the 5e motif form four base pairs which are interrupted by a loop of three nucleotides. In the eukaryotes, the first nucleotide of the loop is an adenosine which is needed for the binding of protein SRP72.

UGU(NR)

The UGU(NR) motif connects helices 3 and 4 in the small (Alu) SRP domain. Fungal SRP RNAs lacking helices 3 and 4 contain the motif within the loop of helix 2. It is important in the binding of the SRP9/14 protein heterodimer as part of an RNA U-turn.

Tertiary

X-ray crystallography, nuclear magnetic resonance (NMR), and cryo-electron microscopy (cryo-EM] have been used to determine the molecular structure of portions of the SRP RNAs from various species. The available PDB structures show the RNA molecule either free or when bound to one or more SRP proteins.

Binding Proteins

One or more SRP proteins bind to the SRP RNA to assemble the functional SRP. The SRP proteins are named according to their approximate molecular mass measured in kilodalton. Most bacterial SRPs are composed of SRP RNA and SRP54 (also named Ffh for "*F*ifty-*f*our *h*omolog"). The Archaeal SRP contains proteins SRP54 and SRP19. In eukaryotes, the SRP RNA combines with the imported SRP proteins SRP9/14, SRP19, and SRP68/72 in a region of the nucleolus. This pre-SRP is transported to the cytosol where it binds to protein SRP54. The molecular structures of the free or SRP RNA-bound proteins SRP9/14, SRP19, or SRP54 are known at high resolution.

SRP9 and SRP14

SRP9 and SRP14 are structurally related and form the SRP9/14 heterodimer which binds to the SRP RNA of the small (Alu) domain.Yeast SRP lacks SRP9 and contains the structurally related protein SRP21. Yeast SRP14 forms a homodimer. SRP9/14 is absent in the SRP of trypanosoma which instead possess a tRNA-like molecule.

SRP19

SRP19 is found in the SRP of eukaryotes and Archaea. Its primary role is in preparing the SRP RNA for the binding of SRP54, SRP68, and SRP72 by properly arranging SRP RNA helices 6 and 8. Yeast SRP contains Sec65p, a larger homolog of SRP19.

SRP54

Protein SRP54 (named Ffh in the bacteria) is an essential component of every SRP. It is composed of three functional domains: the N-terminal (N) domain, the GTPase (G) domain, and the methionine-rich (M) domain.

SRP68 and SRP72

Proteins SRP68 and SRP72 are constituents of the large domain of the eukaryotic SRP. They form a stable SRP68/72 heterodimer. About one third of the human SRP68 protein was shown to bind to the SRP RNA. A relatively small region located near the C-terminus of SRP72 binds to the 5e SRP RNA motif.

Transfer RNA

Transfer RNA (tRNA) is a small RNA molecule (usually about 73-95nucleotides) that transfers a specific active amino acid to a growing polypeptide chain at the ribosomal site of protein synthesis during translation. It has a 3' terminal site for amino acid attachment.

This covalent linkage is catalysed by an aminoacyl tRNA synthetase. It also contains a three base region called the anticodon that can base pair to the corresponding three base codon region on mRNA. Each type of tRNA molecule can be attached to only one type of amino acid, but because the genetic code contains multiple codons that specify the same amino acid, tRNA molecules bearing different anticodons may also carry the same amino acid.

Structure

The structure of tRNA can be decomposed into its primary structure, its secondary structure(usually visualized as the *cloverleaf structure*), and its tertiary structure (all tRNAs have a similar L-shaped 3D structure that allows them to fit into the P and A sites of the ribosome). The cloverleaf structure becomes the 3D L-shaped structure through coaxial stacking of the helices which is a common RNA Tertiary Structure motif.

1. The 5'-terminal phosphate group.
2. The acceptor stem is a 7-base pair (bp) stem made by the base pairing of the 5'-terminal nucleotide with the 3'-terminal nucleotide (which contains the CCA 3'-terminal group used to attach the amino acid). The acceptor stem may contain non-Watson-Crick base pairs.
3. The CCA tail is a cytosine-cytosine-adenine sequence at the 3' end of the tRNA molecule. This sequence is important for the recognition of tRNA by enzymes critical in translation. In prokaryotes, the CCA sequence is transcribed in some tRNA sequences. In most prokaryotic tRNAs and eukaryotic tRNAs, the CCA sequence is added during processing and therefore does not appear in the tRNA gene.
4. The D arm is a 4 bp stem ending in a loop that often contains dihydrouridine.
5. The anticodon arm is a 5-bp stem whose loop contains the anticodon.
6. The T arm is a 5 bp stem containing the sequence TψC where ψ is a pseudouridine.
7. Bases that have been modified, especially by methylation, occur in several positions throughout the tRNA. The first anticodon base, or wobble-position, is sometimes modified to inosine (derived from adenine), pseudouridine (derived from uracil) or lysidine (derived from cytosine).

Anticodon

An anticodon is a unit made up of three nucleotides that correspond to the three bases of the codon on the mRNA. Each tRNA contains a specific anticodon triplet sequence that can base-pair to one or more codons for an amino acid. For example, the codon for lysine is AAA; the anticodon of a lysine tRNA might be UUU. Some anticodons can pair with more than one codon due to a phenomenon known as wobble base pairing. Frequently, the first nucleotide of the anticodon is one of two not found on mRNA: inosine and pseudouridine, which can hydrogen bond to more than one base in the corresponding codon position. In the genetic code, it is common for a single amino acid to be specified by all four third-position possibilities, or at least by both Pyrimidines and Purines; for example, the amino acid glycine is coded for by the codon sequences GGU, GGC, GGA, and GGG. To provide a one-to-one correspondence between tRNA molecules and codons that specify amino acids, 61 types of tRNA molecules would be required per cell. However, many cells contain fewer than 61 types of tRNAs because the wobble base is capable of binding to several, though not necessarily all, of the codons that specify a particular amino acid. A minimum of 31 tRNA are required to translate, unambiguously, all 61 sense codons of the standard genetic code.

Aminoacylation

Aminoacylation is the process of adding an aminoacyl group to a compound. It produces tRNA molecules with their CCA 3' ends covalently linked to an amino acid.

Each tRNA is aminoacylated (or *charged*) with a specific amino acid by an aminoacyl tRNA synthetase. There is normally a single aminoacyl tRNA synthetase for each amino acid, despite the fact that there can be more than one tRNA, and more than one anticodon, for an amino acid. Recognition of the appropriate tRNA by the synthetases is not mediated solely by the anticodon, and the acceptor stem often plays a prominent role.

Reaction

1. amino acid + ATP → aminoacyl-AMP + PPi
2. aminoacyl-AMP + tRNA → aminoacyl-tRNA + AMP.

Sometimes, certain organisms can have one or more aminoacyl tRNA synthetases missing. This leads to mischarging of the tRNA by a chemically related amino acid. The correct amino acid is made by enzymes that modify the mischarged amino acid to the correct one. For example, *Helicobacter pylori* has glutaminyl tRNA synthetase missing. Thus, glutamate tRNA synthetase mischarges tRNA-glutamine (tRNA-Gln) with glutamate. An amidotransferase then converts the acid side chain of the glutamate to the amide, forming the correctly charged gln-tRNA-Gln.

Binding to Ribosome

The ribosome has three binding sites for tRNA molecules: the A (aminoacyl), P (peptidyl), and E (exit) sites. During translation the A site binds an incoming aminoacyl-tRNA as directed by the codon currently occupying this site. This codon specifies the next amino acid to be added to the growing peptide chain. The A site only works after the first aminoacyl-tRNA has attached to the P site. The P-site codon is occupied by peptidyl-tRNA that is a tRNA with multiple amino acids attached as a long chain. The P site is actually the first to bind to aminoacyl tRNA. This tRNA in the P site carries the chain of amino acids that has already been synthesized. The E site is occupied by the empty tRNA as it's about to exit the ribosome.

tRNA Genes

Organisms vary in the number of tRNA genes in their genome. The nematode worm *C. elegans*, a commonly used model organism ingenetics studies, has 29,647 genes in its nuclear genome, of which 620 code for tRNA. The budding yeast *Saccharomyces cerevisiae*has 275 tRNA genes in its genome. In the human genome, which according to current estimates has about 27,161 genes in total, there are about 4,421 non-coding RNA genes, which include tRNA genes. There are 22 mitochondrial tRNA genes; 497 nuclear genes encoding cytoplasmic tRNA molecules and there are 324 tRNA-derived putative pseudogenes.

Cytoplasmic tRNA genes can be grouped into 49 families according to their anticodon features. These genes are found on all chromosomes, except 22 and Y chromosome. High clustering on 6p is observed (140 tRNA genes), as well on 1 chromosome.

tRNA Biogenesis

In eukaryotic cells, tRNAs are transcribed by RNA polymerase III as pre-tRNAs in the nucleus. RNA polymerase III recognizes two internal promoter sequences (A-box B internal promoter) inside tRNA genes. The first promoter begins at nucleotide 8 of mature tRNAs and the second promoter is located 30-60 nucleotides downstream of the first promoter. The transcription terminates after a strech of four or morethymidines.

Pre-tRNAs undergo extensive modifications inside the nucleus. Some pre-tRNAs contain introns; in bacteria these self-splice, whereas in eukaryotes and archaea they are removed by tRNA splicing endonuclease.. The 5' sequence is removed by RNase P,whereas the 3' end is removed by the tRNase Z enzyme. A notable exception is in the archaeon *Nanoarchaeum equitans* which does not possess an RNase P enzyme and has a promoter placed such that transcription starts at the 5' end of the mature tRNA.. The non-templated 3' CCA tail is added by a nucleotidyl transferase. Before tRNAs are exported into the cytoplasm by Los1/Xpo-t, tRNAs are aminoacylated. The order of the processing events is not conserved. For example in yeast, the splicing is not carried out in the nucleus but at the cytoplasmic side of mitochondrial membranes.

History

The existence of tRNA was first hypothesized by Francis Crick, based on the assumption that there must exist an adapter molecule capable of mediating the translation of the RNA alphabet into the protein alphabet. Significant research on structure was conducted in the early 1960s by Alex Rich and Don Caspar, two researchers in Boston, the Jacques Fresco group in Princeton University and a United Kingdom group atKing's College London.

In 1965, a publication by Robert W. Holley reported the primary structure and suggested three secondary structures. The cloverleaf structure was ascertained by several other studies in the following years and was finally confirmed using X-ray crystallography studies in 1974. Two independent groups, Kim Sung-Hou working under Alexander Rich and a British group headed byAaron Klug, published the same crystallography findings within a year.

Transfer-messenger RNA

Transfer-messenger RNA (abbreviated tmRNA, also known as 10Sa RNA and by its genetic name SsrA) is a bacterial RNA molecule with dual tRNA-like and messenger RNA-like properties. The tmRNA forms a ribonucleoprotein complex (tmRNP) together with Small Protein B (SmpB), Elongation Factor Tu (EF-Tu), and ribosomal protein S1. In *trans*-translation, tmRNA and its associated proteins bind to bacterialribosomes which have stalled in the middle of protein biosynthesis, for example when reaching the end of a messenger RNA which has lost its stop codon. The tmRNA is remarkably versatile: it recycles the stalled ribosome, adds a proteolysis-inducing tag to the unfinishedpolypeptide, and facilitates the degradation of the aberrant messenger RNA. In the majority of bacteria these functions are carried out by standard one-piece tmRNAs. In other bacterial species, a permuted *ssrA* gene produces a two-piece tmRNA in which two separate RNA chains are joined by base-pairing.

Discovery of tmRNA and Early Work

tmRNA was first designated 10Sa RNA after a mixed "10S" electrophoretic fraction of *Escherichia coli* RNA was further resolved into tmRNA and the similarly-sized RNase P RNA (10Sb). The presence of pseudouridine in the mixed 10S RNA hinted that tmRNA has modified bases found also intRNA. The similarity at the 3' end of tmRNA to the T stem-loop of tRNA was first recognized upon sequencing *ssrA* from *Mycobacterium tuberculosis*.Subsequent sequence comparison revealed the full tRNA-like domain (TLD) formed by the 5' and 3' ends of tmRNA, including the acceptor stem with elements like those in alanine tRNA that promote its aminoacylation byalanine-tRNA ligase. It also revealed differences from tRNA: the anticodonarm is missing in tmRNA, and the D arm region is a loop without base pairs.

tmRNA Structure

Secondary Structure of the Standard One-piece tmRNAs

The complete *E. coli* tmRNA secondary structure was elucidated by comparative sequence analysis and structural probing. Watson-Crick and G-U base pairswere identified by comparing the bacterial tmRNA sequences using automated computational methods in combination with manual alignment procedures. The accompanying figure shows the base pairing pattern of this prototypical tmRNA, which is organized into 12 phylogenetically supported helices (also called pairings P1 to P12), some divided into helical segments.

A prominent feature of every tmRNA is the conserved tRNA-like domain (TLD), composed of helices 1, 12, and 2a (analogs of the tRNA acceptor stem, T-stem and variable stem, respectively), and containing the 5' monophosphate and alanylatable 3' CCA ends.

The mRNA-like region (MLR) is in standard tmRNA a large loop containing pseudoknots and a coding sequence (CDS) for the tag peptide, marked by the resume codon and the stop codon. The encoded tag peptide (ANDENYALAA in *E. coli*) varies among bacteria, perhaps depending on the set of proteases and adaptors available.

tmRNAs typically contain four pseudoknots, one (pk1) upstream of the tag peptide CDS, and the other three pseudoknots (pk2 to pk4) downstream of the CDS. The pseudoknot regions, although generally conserved, are evolutionarily plasic. For example, in the (one-piece) tmRNAs of cyanobacteria, pk4 is substituted with two tandemly arranged smaller pseudoknots.

This suggests that tmRNA folding outside the TLD can be important, yet the pseudoknot region lacks conserved residues and pseudoknots are among the first structures to be lost as *ssrA* sequences diverge in plastid and endosymbiont lineages. Base pairing in the three-pseudoknot region of *E. coli* tmRNA is disrupted during *trans*-translation.

Two-piece tmRNAs

Circularly permuted *ssrA* has been reported in three major lineages: i) all alphaproteobacteria and the primitive mitochondria of jakobid protists, ii) two disjoint groups of cyanobacteria (*Gloeobacter* and a clade containing *Prochlorococcus* and many *Synechococcus*), and iii) some members of the betaproteobacteria (*Cupriavidus* and some Rhodocyclales).

All produce the same overall two-piece (acceptor and coding pieces) form, equivalent to the standard form nicked downstream of the reading frame. None retain more than two pseudoknots compared to the four (or more) of standard tmRNA.

Alphaproteobacteria have two signature sequences: replacement of the typical T-loop sequence TΨCRANY with GGCRGUA, and the sequence AACAGAA in the large loop of the 3′-terminal pseudoknot. In mitochondria, the MLR has been lost, and a remarkable re-permutation of mitochondrial *ssrA* results in a small one-piece product in *Jakoba libera*.

The cyanobacteria provide the most plausible case for evolution of a permuted gene from a standard gene, due to remarkable sequence similarities between the two gene types as they occur in different *Synechococcus* strains.

tmRNA Processing

Most tmRNAs are transcribed as larger precursors which are processed much like tRNA. Cleavage at the 5′ end is by ribonuclease P.Multiple exonucleases can participate in the processing of the 3′ end of tmRNA, although RNase T and RNase PH are most effective. Depending on the bacterial species, the 3'-CCA is either encoded or added by tRNA nucleotidyltransferase.

Similar processing at internal sites of permuted precursor tmRNA explains its physical splitting into two pieces. The two-piece tmRNAs have two additional ends whose processing must be considered. For alphaproteobacteria, one 5′ end is the unprocessed start site of transcription. The far 3′ end may in some cases be the result of rho-independent termination.

Three-dimensional Structures

High-resolution structures of the complete tmRNA molecules are currently unavailable and may be difficult to obtain due the inherent flexibility of the MLR. In 2007, the crystal structure of the *Thermus thermophilus* TLD bound to the SmpB protein was obtained at 3 Å resolution. This structure shows that SmpB mimics the D stem and the anticodon of a canonical tRNA whereas helical section 2a of tmRNA corresponds to the variable arm of tRNA.

A cryo-electron microscopy study of tmRNA at an early stage of *trans*-translation shows the spatial relationship between the ribosome and the tmRNP (tmRNA bound to the EF-Tu protein). The TLD is located near the GTPase-associated centre in the 50S ribosomal subunit; helix 5 and pseudoknots pk2 to pk4 form an arc around the beak of the 30S ribosomal subunit.

Trans-*translation*

Coding by tmRNA was discovered in 1995 when Simpson and coworkers overexpressed a mouse cytokine in *E. coli*and found several truncated cytokine-derived peptideseach tagged at the carboxyl termini with the same 11-amino acid residue extension (A)ANDENYALAA. With the exception of the N-terminal alanine, which comes from the 3' end of tmRNA itself, this tag sequence was traced to a short open reading frame in *E. coli* tmRNA. Recognizing that the tag peptide confers proteolysis, the *trans*-translation model for tmRNA action was proposed.

While details of the *trans*-translation mechanism are under investigation it is generally agreed that tmRNA first occupies the empty A site of the stalled ribosome. Subsequently, the ribosome moves from the 3' end of the truncated messenger RNA onto the resume codon of the MLR, followed by a slippage-prone stage from where translation continues normally until the in-frame tmRNAstop codon is encountered. *Trans-translation* is essential in some bacterial species, whereas other bacteria require tmRNA to survive when subjected to stressful growth conditions. Depending on the organism, the tag peptide may be recognized by a variety of proteases or protease adapters.

Mobile Genetic Elements and the tmRNA Gene

ssrA is both a target for some mobile DNAs and a passenger on others. It has been found interrupted by three types of mobile elements. By different strategies none of these disrupt gene function: group I introns remove themselves by self-splicing,rickettsial palindromic elements (RPEs) insert in innocuous sites, and integrase-encoding genomic islands split their target *ssrA* yet restore the split-off portion.

Non-chromosomal *ssrA* was first detected in a genomic survey of mycobacteriophages (in 10% of the phages). Other mobile elements including plasmids and genomic islands have been found bearing *ssrA*.

One interesting case is *Rhodobacter sphaeroides*ATCC 17025, whose native tmRNA gene is disrupted by a genomic island; unlike all other genomic islands in tmRNA (or tRNA) genes this island has inactivated the native target gene without restoration, yet compensates by carrying its own tmRNA gene. A very unusual relative of *ssrA* is found in the lytic mycobacteriophage DS6A, that encodes little more that the TLD.

Small Nuclear RNA

Small nuclear ribonucleic acid (snRNA) is a class of small RNA molecules that are found within the nucleus of eukaryotic cells. They are transcribed by RNA polymerase II or RNA polymerase III and are involved in a variety of important processes such as RNA splicing (removal of introns from hnRNA), regulation of transcription factors (7SK RNA) or RNA polymerase II (B2 RNA), and maintaining the telomeres.

They are always associated with specific proteins, and the complexes are referred to as small nuclear ribonucleoproteins (snRNP) or sometimes as snurps. These elements are rich in uridine content.

A large group of snRNAs are known as small nucleolar RNAs (snoRNAs). These are small RNA molecules that play an essential role in RNAbiogenesis and guide chemical modifications of ribosomal RNAs (rRNAs) and other RNA genes (tRNA and snRNAs).

They are located in thenucleolus and the Cajal bodies of eukaryotic cells (the major sites of RNA synthesis).

Small nucleolar RNAs (snoRNAs) are a class of small RNA molecules that primarily guide chemical modifications of other RNAs, mainlyribosomal RNAs, transfer RNAs and small nuclear RNAs. There are two main classes of snoRNA, the C/D box snoRNAs which are associated with methylation, and the H/ACA box snoRNAs which are associated with pseudouridylation. snoRNAs are commonly referred to as guide RNAs but should not be confused with the guide RNAs that direct RNA editing in trypanosomes.

snoRNA Guided Modifications

After transcription, nascent rRNA molecules (termed pre-rRNA) are required to undergo a series of processing steps in order to generate the mature rRNA molecule. Prior to cleavage by exo- and endonucleases the pre-rRNA undergoes a complex pattern of nucleoside modifications. These include methylations and pseudouridylations, guided by snoRNAs.

- Methylation is the attachment or substitution of a methyl group onto various substrates. The rRNA of humans contain approximately 115 methyl group modifications. The majority of these are 2'O-ribose-methylations (where the methyl group is attached to the ribose group).
- Pseudouridylation is the conversion (isomerisation) of the nucleoside uridine to a different isomeric form pseudouridine(ψ). Mature human rRNAs contain approximately 95 ψ modifications.

Each snoRNA molecule acts as a guide for only one (or two) individual modifications in a target RNA. In order to carry out modification, each snoRNA associates with at least four protein molecules in an RNA/protein complex referred to as a small nucleolar ribonucleoprotein (snoRNP). The proteins associated with each RNA depend on the type of snoRNA molecule. The snoRNA molecule contains an antisense element (a stretch of 10-20 nucleotides) which are base complementary to the sequence surrounding the base (nucleotide) targeted for modification in the pre-RNA molecule. This enables the snoRNP to recognise and bind to the target RNA. Once the snoRNP has bound to the target site the associated proteins are in the correct physical location to catalyse the chemical modification of the target base.

snoRNA Guide Families

The two different types of rRNA modification (methylation and pseudouridylation) are directed by two different families of snoRNPs. These families of snoRNAs are referred to as antisense C/D box and H/ACA box snoRNAs based on the presence of conserved sequence motifs in the snoRNA. There are exceptions but as a general rule C/D box members guide methylation and H/ACA members guide pseudouridylation. The members of each family may vary in biogenesis, structure and function but each family is classified by the following generalised characteristics.

C/D Box

C/D box snoRNAs contain two short conserved sequence motifs, C (UGAUGA) and D (CUGA) located near the 5' and 3' ends of the snoRNA respectively. Short regions (~ 5 nucleotides) located upstreamof the C box and downstream of the D box are usually base complementary and form a stem-box structure which brings the C and D box motifs into close proximity. This stem-box structure has been shown to be essential for correct snoRNA synthesis and nucleolar localization.

Many C/D box snoRNA also contain an additional less well conserved copy of the C and D motifs (referred to as C' and D') located in the central portion of the snoRNA molecule. A conserved region of 10-21 nucleotides upstream of the D box is complementary to the methylation site of the target RNA and enables the snoRNA to form an RNA duplex with the RNA. The nucleotide to be modified in the target RNA is usually located at the 5th position upstream from the D box (or D' box). C/D box snoRNAs associate with four evolutionary conserved and essential proteins (Fibrillarin (Nop1p), Nop56p, Nop58p, and Snu13 (in eukaryotes; its archaeal homolog is L7Ae)) which make up the core C/D box snoRNP.

There exists a eukaryotic C/D box snoRNA (snoRNA U3) that has not been shown to guide 2'-*O*-methylation. Instead, it functions in rRNA processing by directing pre-rRNA cleavage.

H/ACA Box

H/ACA box snoRNAs have a common secondary structure consisting of a two hairpins and two single stranded regions termed a hairpin-hinge-hairpin-tail structure. H/ACA snoRNAs also contain conserved sequence motifs known as H box (consensus ANANNA) and the ACA box (ACA). Both motifs are usually located in the single stranded regions of the secondary structure. The H motif is located in the hinge and the ACA motif is located in the tail region, 3 nucleotides from the 3' end of the sequence.

The hairpin regions contain internal bulges known as recognition loops in which the antisense guide sequences (bases complementary to the target sequence) are located. This recognition sequence is bipartite (constructed from the two different arms of the loop region) and forms complex pseudo-knots with the target RNA. H/ACA box snoRNAs associate with four evolutionary conserved and essential proteins (dyskerin (Cbf5p), Gar1p, Nhp2p and Nop10p) which make up the core of the H/ACA box snoRNP. However, in lower eukaryotic cells such as trypanosomes, similar RNAs exist in the form of single hairpin structure and an AGA box instead of ACA box at the 3' end of the RNA.

The RNA component of human telomerase (hTERC) contains an H/ACA domain for pre-RNP formation and nucleolar localization of the telomerase RNP itself. Importantly, the H/ACA snoRNP has been implicated in the rare genetic disease dyskeratosis congenita (DKC) due to its affiliation with human telomerase. Mutations in the protein component of the H/ACA snoRNP result in a reduction in physiological TERC levels. This has been strongly correlated with the pathology behind DKC which seems to be primarily a disease of poor telomeremaintenance.

Composite H/ACA and C/D Box

An unusual guide snoRNA U85 was identified that functions in both 2'-O-ribose methylation and pseudouridylation of small nuclear RNA(snRNA) U5. This composite snoRNA contains both C/D and H/ACA box domains and associates with the proteins specific to each class of snoRNA (fibrillaring and Gar1p respectively. More composite snoRNAs have now been characterised.

These composite snoRNAs have been found to accumulate in a subnuclear organelle called the Cajal body and are referred to as Cajal body specific RNAs. This is in contrast to the majority of C/D box or H/ACA box snoRNAs which localise to the nucleolus. These Cajal body specific RNAs and are proposed to be involved in the modification of RNA polymerase II transcribed spliceosomal RNAs U1, U2, U4, U5 and U12. Not all snoRNAs that have been localised to Cajal bodies are composite C/D and H/ACA box snoRNAs.

Orphan snoRNAs

The targets for newly identified snoRNAs are predicted on the basis of sequence complementarity between putative target RNAs and the antisense elements or recognition loops in the snoRNA sequence. However, there are an increasing number of 'orphan' guides without any known RNA targets, which suggests that there might be more proteins or transcripts involved in rRNA than previously and/or that some snoRNAs have different functions not concerning rRNA.

There is evidence that some of these orphan snoRNAs regulate alternatively spliced transcripts. For example, it appears that the C/D box snoRNA SNORD115 regulates the alternative splicing of the serotonin 2C receptor mRNA via a conserved region of complementarity. Another C/D box snoRNA, SNORD116, that resides in the same cluster as SNORD115 has been predicted to have 23 possible targets within protein coding genes using a bioinformatic approach. Of these a large fraction were found to be alternatively spliced, suggesting a role of SNORD116 in the regulation of alternative splicing.

Target Modifications

The precise effect of the methylation and pseudouridylation modifications on the function of the mature RNAs is not yet known. The modifications do not appear to be essential but are known to subtly enhance the RNA folding and interaction with ribosomal proteins. In support of their importance, target site modifications are exclusively located within conserved and functionally important domains of the mature RNA and are commonly conserved amongst distant eukaryotes.

1. 2'-O-methylated ribose causes an increase in the 3'-endo conformation
2. Pseudouridine (psi/ψ) adds another option for H-bonding.
3. Heavily methylated RNA is protected from hydrolysis. rRNA acts as a ribozyme by catalyzing its own hydrolysis and splicing.

Genomic Organisation

The majority of vertebrate snoRNA genes are encoded in the introns of proteins involved in ribosome synthesis or translation, and are synthesized by RNA polymerase II, but snoRNAs can also be transcribed from their own promoters by RNA polymerase II or III.

Imprinted Loci

In the human genome there are at least two examples where C/D box snoRNAs are found in tandem repeats within imprinted loci. These two loci (14q32 on chromosome 14 and 15q11q13 on chromosome 15) have been extensively characterised and in both regions multiple snoRNAs have been found located within introns in clusters of closely related copies.

In 15q11q13, five different snoRNAs have been identified (SNORD64, SNORD107, SNORD108, SNORD109 (two copies), SNORD116 (29 copies) and SNORD115 (48 copies). Loss of the 29 copies of SNORD116 from this region has been identified as a cause of Prader-Willi syndrome whereas gain of additional copies of SNORD115 has been linked to autism.

Region 14q32 contains repeats of two snoRNAs SNORD113 (9 copies) and SNORD114 (31 copies) within the introns of a tissue-specific ncRNA transcript (MEG8). The 14q32 domain has been shown to share common genomic features with the imprinted 15q11-q13 loci and a possible role for tandem repeats of C/D box snoRNAs in the evolution or mechanism of imprinted loci has been suggested.

Other Functions of snoRNA

snoRNAs can function as miRNAs. It has been shown that human ACA45 is a *bona fide* snoRNA that can be processed into a 21nucleotides long mature miRNA by the RNAse III family endoribonuclease dicer. This snoRNA product has previously been identified asmmu-miR-1839 and was shown to be processed independent of the other miRNA generating endoribonuclease drosha. Bioinformaticalanalyses have revealed that putatively snoRNA-derived, miRNA-like fragments occur in different organisms. Recently, it has been found that snoRNAs can have functions not related to rRNA. One such function is the regulation of alternative splicingof the *trans* gene transcript, which is done by the snoRNA HBII-52, which is also known as SNORD115. Recent discovery also show, the existence of snoRNA, microRNA, piRNA characteristics in a novel non-coding RNA: x-ncRNA and its biological implication in *Homo sapiens*.

SmY RNA

SmY ribonucleic acids (SmY RNAs) are a family of small nuclear RNAsfound in some species of nematode worms. They are thought to be involved in mRNA trans-splicing.

SmY RNAs are about 70-90 nucleotides long and share a commonsecondary structure, with two stem-loops flanking a consensus binding sitefor Sm protein. Sm protein is a shared component of spliceosomalsnRNPs.

mY RNAs have been found in nematodes of class Chromadorea, which includes the most commonly studied nematodes (such as *Caenorhabditis*, *Pristionchus*, and *Ascaris*), but not in the more distantly related *Trichinella spiralis* in class Dorylaimia. The number of SmY genes in each species varies, with most *Caenorhabditis* and *Pristionchus* species having 10-26 related paralogous copies, while other nematodes have 1-5.

Discovery

The first SmY RNA was discovered in 1996 in purified *Ascaris lumbricoides*spliceosome preparations, as was another called SmX RNA that is not detectably homologous to SmY. Twelve SmY homologs were identifiedcomputationally in *Caenorhabditis elegans*, and ten in *Caenorhabditis briggsae*. Several transcripts from these SmY genes were cloned andsequenced in a systematic survey of small non-coding RNA transcripts in*C. elegans*.

A systematic survey of 2,2,7-trimethylguanosine (TMG) 5' cappedtranscripts in *C.elegans* using anti TMG antibodies identified two TMG capped SmY transcripts. Sequence analysis of the potential Sm binding sites in these transcripts indicated the SmY, U5 snRNA, U3 snoRNA and the spliced leader RNAs transcripts (SL1 and SL2) all contain a very similar consensus SM binding sequence ($AAU_4–_5GGA$). The predicted SM binding sites identified in the U1, U2 and U4 snRNA transcripts varied from this consensus.

Function

In *C. elegans*, SmY RNAs copurify with spliceosome and with Sm, SL75p, and SL26p proteins, while the better-characterized *C. elegans*SL1 trans-splicing snRNA copurifies in a complex with Sm, SL75p, and SL21p (a paralog of SL26p). Loss of function of either SL21p or SL26p individually causes only a weak cold-sensitive phenotype, whereas knockdown of both is lethal, as is a SL75p knockdown. Based on these results, the SmY RNAs are believed to have a function in trans-splicing.

Small Cajal Body-specific RNA

Small Cajal body-specific RNAs (scaRNAs) are a class of small nucleolar RNAs (snoRNAs) which specifically localise to the Cajal body, a nuclear organelle involved in the biogenesis of small nuclear ribonucleoproteins (snRNPs or snurps). ScaRNAs guide the modification (methylation and pseudouridylation) of RNA polymerase IItranscribed spliceosomal RNAs U1, U2, U4, U5 and U12.

The first scaRNA identified was U85. It is unlike typical snoRNAs in that it is a composite C/D box and H/ACA box snoRNAs and can guide both pseudouridylation and 2'-O-methylation. Not all scaRNAs are composite C/D and H/ACA box snoRNA and most scaRNAs are structurally and functionally indistinguishable from snoRNAs directing ribosomal RNA (rRNA) modification in the nucleolus.

Small Cajal Body-specific RNA 1

In molecular biology, Small Cajal body specific RNA 1 (also known as *scaRNA1* or*ACA35*) is a small nucleolar RNA found in Cajal bodies and believed to be involved in the pseudouridylation of U2 spliceosomal RNA.

scaRNA1 is a non-coding RNA, which are functional products of gene but nottranslated into a protein. Such RNA molecules usually contain important secondary structure or ligand binding motifs and are involved in many important biological processes in the cell. More specifically, scaRNAs are a class of small nuclear RNAsthat localise to the Cajal bodies and guide the modification of RNA polymerase IItranscribed spliceosomal RNAs U1, U2, U4, U5 and U12.

scaRNA1 belongs to the H/ACA box class of guide RNAs, as it has the predicted hairpin-hinge-hairpin-tail structure, conserved H/ACA-box motifs and is found associated with GAR1 protein. It is predicted to guide pseudouridylation of the U2 spliceosomal RNA at residue U89.

Guide RNA

Guide RNAs (aka gRNA) are the RNAs that guide the insertion or deletion of uridine residues into mitochondrial mRNAs in kinetoplastidprotists in a process known as RNA editing.

Overview of gRNA-directed Editing

Trypanosomatid protists and other kinetoplastids have a novel post-transcriptional mitochondrial RNA modification process known as "RNA editing". The mitochondrial genome in these cells consists of 20-50 maxicircles that encode genes and "cryptogenes" (and some gRNAs) and 10-20,000 minicircles that encode gRNAs. All of these molecules are catenated into a giant network of DNA that is situated at the base of the flagellum in the inner compartment of the single mitochondrion.

A majority of the maxicircle transcripts can not be translated into proteins due to multiple frameshifts in the sequences. These frameshifts are corrected after transcription by the insertion and deletion of uridine residues at precise sites which create an open reading frame that is translated into a mitochondrial protein homologous to mitochondrial proteins from other cells. The insertions and deletions are mediated by short guide RNA (gRNAs) which encode the editing information in the form of complementary sequences (allowing GU as well as GC base pairs). The gRNAs are transcribed from both the maxicircles and the minicircles.

The presence of two genomes in the mitochondrion, one of which contains sequence information that corrects errors in the other genome, is novel. Editing proceeds generally 3' to 5' on the mRNA. The initial editing event occurs when a gRNA forms an RNA duplex with a complementary mRNA sequence just downstream of the editing site. This then recruits a number of ribonucleoprotein complexes that direct the precise insertions and deletions of uridine residues, thereby extending the duplex upstream. The adjacent upstream editing site is then modified in the same manner. A single gRNA usually encodes the information for several editing sites (an editing "block"), the editing of which produces a complete gRNA/mRNA duplex.

In the case of "pan-edited" mRNAs, the duplex unwinds and another gRNA then forms a duplex with the edited mRNA sequence and initiates another round of editing. The overlapping gRNAs form an editing "domain". In some genes there are multiple editing domains. The extent of editing for any particular gene varies between trypanosomatid species.

The variation consists of the loss of editing at the 3' side, probably due to the loss of minicircle sequence classes that encode specific gRNAs. A retroposition model has been proposed to account for the partial, and in some cases, complete, loss of editing in evolution. Loss of editing is lethal in most cases, although losses have been seen in old laboratory strains. The maintenance of editing over the long evolutionary history of these ancient protists implies the presence of a selective advantage, the exact nature of which is still uncertain.

It is not clear why trypanosomatids utilize such an elaborate mechanism to produce mRNAs. It may have originated in the early mitochondrion of the ancestor of the kintoplastid protist lineage, since it is present in the bodonids which are ancestral to the trypanosomatids, and may not be present in the euglenoids, which branched from the same common ancestor as the kinetoplastids.

In the protozoan *Leishmania tarentolae*, 12 of the 18 mitochondrial genes are edited using this process. One such gene is Cyb. The mRNA is actually edited twice in succession. For the first edit, the relevant sequence on the mRNA is:

mRNA 5' AAAGAAAAGGCUUUAACUUCAGGUUGU 3'

The 3' end is used to anchor the gRNA (gCyb-I gRNA in this case) by basepairing (some G/U pairs are used). The 5' end does not exactly match and one of three specific endonucleases cleaves the mRNA at the mismatch site.

gRNA 3'
AAUAAUAAAUUUUUAAAUAUAAUAGAAAAUUGAAGUUCAGUA 5'

mRNA 5' A A AGAAA A G G C UUUAACUUCAGGUUGU 3'

The mRNA is now "repaired" by adding U's at each editing site in succession, giving the sequence

gRNA 3'
AAUAAUAAAUUUUUAAAUAUAAUAGAAAAUUGAAGUUCAGUA 5'

mRNA 5'
UUAUUAUUUAGAAAUUUAUGUUGUCUUUUAACUUCAGGUUGU 3'

This particular gene has two overlapping gRNA editing sites. The 5' end of this section is the 3' anchor for another gRNA (gCyb-II gRNA).

RNase P

Ribonuclease P (RNase P) is a type of Ribonuclease which cleaves RNA. RNase P is unique from other RNases in that it is a ribozyme – a ribonucleic acid that acts as a catalyst in the same way that a protein based enzyme would. Its function is to cleave off an extra, or precursor, sequence of RNA on tRNA molecules. Further RNase P is one of two known multiple turnover ribozymes in nature (the other being the ribosome), the discovery of which earned Sidney Altmanand Thomas Cech the Nobel Prize in Chemistry in 1989.

In fact, Sidney Altman discovered the existence of precursor tRNA with flanking sequences and was the first to characterize RNase P and its activity in processing of the 5' leader sequence of precursor tRNA back in the 70's. Recent findings also reveal that RNase P has a new function. It has been shown that human nuclear RNase P is required for the normal and efficient transcription of various small noncoding RNA genes, such as tRNA, 5S rRNA, SRP RNA and U6 snRNA genes, which are transcribed byRNA polymerase III, one of three major nuclear RNA polymerases in human cells.

In Bacteria

In bacteria, such as *E. coli*, RNase P has two components: an RNA chain, called M1 RNA, and a polypeptide chain, or protein, called C5 protein,. In vivo, both components are necessary for the ribozyme to function properly, but in vitro, the M1 RNA can act alone as a catalyst. The primary role of the C5 protein is to enhance the substrate binding affinity and the catalytic rate of the M1 RNA enzyme probably by increasing the metal ion affinity in the active site. The crystal structure of bacterial RNase P RNA has been recently resolved, revealing its flat surface that is generated by a number of coaxially stacked helical domains connected by local and long-range contacts. This flat surface facilitates binding and cleavage of precursor tRNA substrates.

Bacterial RNase P class A and B

Ribonuclease P (RNase P) is a ubiquitous endoribonuclease, found in archaea, bacteria and eukarya as well as chloroplasts and mitochondria. Its best characterised activity is the generation of mature 5'-ends of tRNAs by cleaving the 5'-leader elements of precursor-tRNAs. Cellular RNase Ps are ribonucleoproteins. RNA from bacterial RNase Ps retains its catalytic activity in the absence of the protein subunit, i.e. it is a ribozyme. Isolated eukaryotic and archaeal RNase P RNA has not been shown to retain its catalytic function, but is still essential for the catalytic activity of the holoenzyme. Although the archaeal and eukaryotic holoenzymes have a much greater protein content than the bacterial ones, the RNA cores from all the three lineages are homologous — helices corresponding to P1, P2, P3, P4, and P10/11 are common to all cellular RNase P RNAs. Yet, there is considerable sequence variation, particularly among the eukaryotic RNAs.

In Archaea

In archaea, RNase P ribonucleoproteins consist of 4-5 protein subunits that are associated with RNA. As revealed by in vitro reconstitution experiments these protein subunits are individually dispensable for tRNA processing that is essentially mediated by the RNA component. The structures of protein subunits of archaeal RNase P have been resolved by x-ray crystallography and NMR, thus revealing new protein domains and folding fundamental for function.

Isolated eukaryotic and archaeal RNase P RNA has not been shown to retain its catalytic function, but is still essential for the catalytic activity of the holoenzyme. Although the archaeal and eukaryotic holoenzymes have a much greater protein content than the bacterial ones, the RNA cores from all the three lineages are homologous — helices corresponding to P1, P2, P3, P4, and P10/11 are common to all cellular RNase P RNAs. Yet, there is considerable sequence variation, particularly among the eukaryotic RNAs.

It has recently been argued that the archaebacteriium *Nanoarchaeum equitans* does not possess RNase P. Computational and experimental studies failed to find evidence for its existence. In this organism the tRNA promoter is close to the tRNA gene and it is thought that transcription starts at the first base of the tRNA thus removing the requirement for RNase P.

In Eukaryotes

In eukaryotes, such as humans and yeast, most RNase P consists of an RNA chain that is structurally similar to that found in bacteria as well as nine to ten associated proteins (as opposed to the single bacterial RNase P protein, C5). Five of these protein subunits exhibit homology to archaeal counterparts. These protein subunits of RNase P are shared with RNase MRP, a catalytic ribonucleoprotein involved in processing of ribosomal RNA in the nucleolus. RNase P from eukaryotes was only recently demonstrated to be a ribozyme. Accordingly, the numerous protein subunits of eucaryal RNase P have a minor contribution to tRNA processing per se, while they seem to be essential for the function of RNase P and RNase MRP in other biological settings, such as gene transcription and the cell cycle. It has recently been discovered that human mitochondrial RNase P is a proteinand does not contain RNA.

RNase MRP

RNase MRP is an enzymatically active ribonucleoprotein with two distinct roles ineukaryotes. In mitochondria it plays a direct role in the initiation of mitochondrial DNA replication. In the nucleus it is involved in precursor rRNA processing, where it cleavesthe internal transcribed spacer 1 between 18S and 5.8S rRNAs. Despite distinct functions, RNase MRP has been shown to be evolutionarily related to RNase P. Like eukaryotic RNase P, RNase MRP is not catalytically active without associated protein subunits.

Y RNA

Y RNAs are small non-coding RNA components of the Ro ribonucleoprotein particle (Ro RNP). The Ro RNP was first identified by Lerner *et al.*. as a target of autoimmune antibodies in patients with systemic lupus erythematosus.

Function

Two functions have been described for Y RNAs in the literature: In one line of evidence, Y RNA appears to function as a repressor of Ro. In its free state, Ro binds to a variety of misfolded RNAs including misfolded 5S rRNAs, and is thought to act as some sort of quality control mechanism. Crystal structures of Ro complexed either with Y RNA or another RNA showed that Ro binds single-stranded 3' ends of RNAs relatively nonspecifically, whereas Y RNA binds specifically at a second site that regulates access of other RNAs.

In*Deinococcus*, free Ro has also been shown to function in 23S rRNA maturation. In*Deinococcus*, mutants lacking Y RNA are viable, and Y RNA appears to be unstable except when complexed with Ro. Secondly, it has been found that human Y RNAs are functionally required for DNA replication.

Biochemical fractionation and reconstitution experiments have established a functional requirement of human Y RNAs for chromosomal DNA replication in isolated vertebrate cell nuclei *in vitro*.Specific degradation of human Y RNAs *in vitro*, or in intact cells *in vivo*, have led to an inhibition of chromosomal DNA replication. Interestingly, mutant human Y RNAs lacking the conserved binding site for Ro60 protein still support DNA replication, indicating that binding to Ro protein and promoting DNA replication are two separable functions of Y RNAs. Another study, has shown that Y RNAs are overexpressed in human tumours and required for cell proliferation.

Structure

These small RNAs are predicted to fold into a conserved stem formed by the RNA's 3' and 5' ends and characterized by a single bulgedcytosine, these are the known requirements for Ro binding.

Species Distribution

Presumptive Y RNA and Ro protein homologs have been found in eukaryotes and bacteria. Humans appear to have four Y RNAs, named hY1, hY3, hY4 and hY5 and also a large number of pseudogenes. *C. elegans* has one, named CeY RNA and a large number of sbRNAsthat are postulated to also be Y RNA homologues.

The radiation-resistant bacterium *Deinococcus radiodurans* encodes a homolog of Ro called rsr ("Ro sixty related"), and at least four small RNAs accumulate in *Deinococcus* under conditions where rsr expression is induced (UV irradiation); one of these RNAs appears to be a Y RNA homolog.

Telomerase RNA Component

Telomerase RNA component, also known as TERC, is an RNA gene found ineukaryotes, that is a component of telomerase used to extend telomeres. Telomerase RNAs differ greatly in sequence and structure between vertebrates, ciliates and yeasts, but they share a 5' pseudoknot structure close to the template sequence. The vertebrate telomerase RNAs have a 3' H/ACA snoRNA-like domain.

Function

Telomerase is a ribonucleoprotein polymerase that maintains telomere ends by addition of the telomere repeat TTAGGG, this repeat does vary across eukaryotes. The enzyme consists of a protein component with reverse transcriptase activity, and an RNA component, encoded by this gene, that serves as a template for the telomere repeat. Telomerase expression plays a role in cellular senescence, as it is normally repressed in postnatal somatic cells resulting in progressive shortening of telomeres. Deregulation of telomerase expression in somatic cells may be involved in oncogenesis. Studies in mouse suggest that telomerase also participates in chromosomal repair, since de novo synthesis of telomere repeats may occur at double-stranded breaks. Homologs of TERC can also be found in the Gallid herpes viruses.

Clinical Significance

Mutations in this gene cause autosomal dominant dyskeratosis congenita, and may also be associated with some cases of aplastic anemia.

Antisense RNA

Antisense RNA is a single-stranded RNA that is complementary to a messenger RNA (mRNA) strand transcribed within a cell. Antisense RNA may be introduced into a cell to inhibit translation of a complementary mRNA by base pairing to it and physically obstructing the translation machinery. This effect is therefore stoichiometric. An example of naturally occurring mRNA antisense mechanism is thehok/sok system of the E.coli R1 plasmid. Antisense RNA has long been thought of as a promising technique for disease therapy; the only such case to have reached the market is the drug fomivirsen. One commentator has characterized antisense RNA as one of "dozens of technologies that are gorgeous in concept, but exasperating in [commercialization]". Generally, antisense RNA still lack effective design, biological activity, and efficient route of administration.

Historically, the effects of antisense RNA have often been confused with the effects of RNA interference (RNAi), a related process in which*double-stranded* RNA fragments called small interfering RNAs trigger catalytically mediated gene silencing, most typically by targeting theRNA-induced silencing complex (RISC) to bind to and degrade the mRNA. Attempts to genetically engineer transgenic plants to express antisense RNA instead activate the RNAi pathway, although the processes result in differing magnitudes of the same downstream effect, gene silencing. Well-known examples include the Flavr Savr tomato and two cultivars of ringspot-resistant papaya.

Transcription of longer cis-antisense transcripts is a common phenomenon in the mammalian transcriptome. Although the function of some cases have been described, such as the Zeb2/Sip1 antisense RNA, no general function has been elucidated. In the case of Zeb2/Sip1, the antisense noncoding RNA is opposite the 5' splice site of an intron in the 5'UTR of the Zeb2 mRNA. Expression of the antisense ncRNA prevents splicing of an intron that contains a ribosome entry site necessary for efficient expression of the Zeb2 protein. Transcription of long antisense ncRNAs is often concordant with the associated protein-coding gene, but more detailed studies have revealed that the relative expression patterns of the mRNA and antisense ncRNA are complex.

Cis-natural Antisense Transcript

Natural antisense transcripts (NATs) are a group of RNAs encoded within a cell that have transcript complementarity to other RNA transcripts. They have been identified in multiple eukaryotes, including humans, mice, yeast and *Arabidopsis thaliana*. This class of RNAs includes both protein-coding and non-coding RNAs. Current evidence has suggested a variety of regulatory roles for NATs, such asRNA interference (RNAi), alternative splicing, genomic imprinting, and X-chromosome inactivation. NATs are broadly grouped into two categories based on whether they act in cis or in trans. Trans-NATs are transcribed from a different location than their targets and usually have complementarity to multiple transcripts with some mismatches. MicroRNAs (miRNA) are an example of trans-NATs that can target multiple transcripts with a few mismatches. Cis-natural antisense transcripts (cis-NATs) on the other hand are transcribed from the same genomic locus as their target but from the opposite DNA strand and form perfect pairs.

Orientation of Cis-NATs

Cis-NATs have a variety of orientations and differing lengths of overlap between pairs. There have been five identified orientations for cis-NATs to date. The most common orientation is head-to-head, where the 5' ends of both transcripts align together. This orientation would result in the greatest knockdown of gene expression if transcriptional collision is the reason for transcript inhibition.

There is however some studies that have suggested that tail-to-tail orientations are the most common NAT pairs. Others such as tail to tail, overlapping, nearby head-to- head, and nearby tail-to-tail are less frequently encountered. Completely overlapping NATs involve the antisense gene being located completely over top of each other. Nearby head-to-head and tail-to-tail orientations are physically discrete from each other but are located very close to each other. Current evidence suggests that there is an overrepresentation of NAT pairs in genes that have catalytic activity. There may be something about these genes in particular that makes them more prone to this type of regulation.

Identification Approach

Identification of NATs in whole genomes is possible due to the large collection of sequence data available from multiple organisms. *In silico*methods for detecting NATs suffer from several shortcomings depending on the source of sequence information. Studies that use mRNAhave sequences whose orientations are known, but the amount of mRNA sequence information available is small. Predicted gene models using algorithms trained to look for genes gives an increased coverage of the genome at the cost of confidence in the identified gene. Another resource is the extensive expressed sequence tag (EST) libraries but these small sequences must first be assigned an orientation before useful information can be extracted from them. Some studies have utilized special sequence information in the ESTs such as the poly(A) signal, poly(A) tail, and splicing sites to both filter the ESTs and to give them the correct transcriptional orientation. Combinations of the different sequence sources attempts to maximize coverage as well as maintain integrity in the data.

Pairs of NATs are identified when they form overlapping clusters. There is variability in the cut-off values used in different studies but generally ~20 nucleotides of sequence overlap is considered the minimum for transcripts to be considered and overlapping cluster. Also, transcripts must map to only one other mRNA molecule in order for it to be considered a NAT pair. Currently there are a variety of web and software resources that can be used to look for antisense pairs. The NATsdb or Natural Antisense Transcript database is a rich tool for searching for antisense pairs from multiple organisms.

Mechanisms

Molecular mechanisms behind the regulatory role of cis-NATs are not currently well understood. Three models have been proposed to explain the regulatory effects that cis-NATs have on gene expression. The first model attributes that base pairing between the cis-NAT and its complementary transcript result in a knockdown of mRNA expression. The assumption of this model is that there will be a precise alignment of at least 6 base pairs between the cis-NAT pair to make double stranded RNA. Epigenetic modifications likeDNA methylation and post-translational modification of core histones form the basis of the second model. Although it is not yet clearly understood, it is thought that the reverse transcript guides methylation complexes and/or histone-modifying complexes to the promoterregions of the sense transcript and cause an inhibition of expression from the gene. Currently it is not known what attributes of cis-NATs are crucial for the epigenetic model of regulation. The final proposed model that has gained favour due to recent experimental evidence is the transcriptional collision model. During the process of transcription of cis-NATs, the transcriptional complexes assemble in the promoter regions of the gene. RNA polymerases will then begin transcribing the gene at the transcription initiation site laying down nucleotides in a 5' to 3' direction. In the areas of overlap between the cis-NATs the RNA polymerases will collide and stop at the crash site. Transcription is inhibited because RNA polymerases prematurely stop and their incomplete transcripts get degraded.

Importance

Regulation of many biological processes such as development, metabolism and many others requires a careful co-ordination between many different genes; this is usually referred to as a gene regulatory network. A flurry of interest in gene regulatory networks has been sparked by the advent of sequenced genomes of multiple organisms. The next step is to use this information to figure out how genes work together and not just in isolation.

During the processes of mammalian development, there is an inactivation of the extra X-chromosome in females. It has been shown that a NAT pair called Xist and Tsix are involved in the hypermethylation of the chromosome. As much as 20-30% of mammalian genes have been shown to be the targets of miRNAs, which highlights the importance of these molecules as regulators across a wide number of genes. Evolutionary reasons for utilizing RNA for regulation of genes may be that it is less costly and faster than synthesizing proteins not needed by the cell. This could have had a selective advantage for early eukaryotes with this type of transcriptional regulation.

Cis-NATs and Disease

Antisense transcription might contribute to disease through chromosomal changes that result in the production of aberrant antisense transcripts. A documented case of cis-NATs being involved in human disease comes from an inherited form of α-thalassemia where there is silencing of the hemoglobin α-2 gene through the action of a cis-NAT. It is thought that in malignant cancer cells with activated transposable elements creates a large amount of transcriptional noise. It is likely that aberrant antisense RNA transcripts resulting from this transcriptional noise may cause stochastic methylation of CpG islands associated withoncogenes and tumour suppressor genes. This inhibition would further progress the malignancy of the cells since they lose key regulator genes. By looking at upregulated antisense transcripts in tumour cells, researchers and look for more candidate tumour suppressor genes. Also, aberrant cis-NATs have been implicated in neurological diseases such as Parkinson's disease.

CRISPR

CRISPRs (Clustered Regularly Interspaced ShortPalindromic Repeats) are sites containing multiple short direct repeats found in the genomes of approximately 40% of bacteria and 90% of archaea. CRISPR functions as a prokaryotic immune system, in that it confers resistance to exogenous genetic elements, such as plasmids and phage. The CRISPR system provides a form of acquired immunity against exogenous DNA by integrating foreign DNA sequences between the repeat sequences in the CRISPR locus of the prokaryote. The integrated DNA sequence is then utilized by the CRISPR system to recognize and silence exogenous genetic elements in a manner analogous to RNAi in eukaryotic organisms.

Discovery and Properties

The existence of the DNA fragments, which are known as CRISPR today, was discovered in 1987 in the bacterium *E. coli*. In 2000 it was announced that there exist similar structures in the genome of many different prokaryotes, describing the Short Regularly Spaced Repeats (SRSR) family. SRSR were renamed as CRISPR in 2002, when a group of genes discovered near the locus of CRISPR repeats called CAS (CRISPR-associated) were identified.

CRISPR Spacers and Repeats

CRISPR repeats range in size from 24 to 48 base pairs. They usually show some dyad symmetry but are not truly palindromic, implying the presence of a conserved secondary structure such as a hairpin. The repeats are separated by unique sequences or "spacers" of similar length. The spacers are usually unique in a genome and have been found to match sequences in phage genomes. These spacers can be acquired from phage and subsequently help protect the cell from subsequent infections. An active CRISPR repeat array can evolve rapidly in response to phage infections. Different strains of the same species of bacterium can be distinguished by differences in their CRISPR arrays by a technique called spoligotyping.

CRISPR Mechanism

Exogenous DNA is apparently processed by proteins encoded by some of the CRISPR-associated (cas) genes into small elements (of ~30bp in length), which are then somehow inserted into the CRISPR locus near the leader sequence. RNAs from the CRISPR loci are constituitively expressed and are processed by Cas proteins to small RNAs composed of individual exogenously-derived sequence elements with some flanking repeat sequence. The RNAs guide other Cas proteins to silence exogenous genetic elements at the RNA or DNA level.

CAS Genes

More than forty different Cas protein families have been described. The cas genes are tightly associated with CRISPRs. The most important of the Cas proteins appears to be Cas1, which is ubiquitous in the CRISPR/Cas system. Particular combinations of cas genes are found together, along with characteristic subclasses of CRISPR repeat sequences and corresponding subfamilies of the Cas1 protein. These combinations appear to represent distinct CRISPR/Cas subtypes. Several different subtypes may occur in a single genome. The sporadic distribution of the CRISPR/Cas subtypes suggests that the system is subject to horizontal gene transfer during microbial evolution. The Cse (Cas subtype *E. coli*) proteins (called CasA-E in *E. coli*) form a functional complex, Cascade, that processes CRISPR RNA transcripts into spacer-repeat units that are retained by Cascade. In other prokaryotes, Cas6 processes the CRISPR transcripts. Interestingly, CRISPR-based phage inactivation in *E. coli* requires Cascade and Cas3, but not Cas1 and Cas2. The Cmr (Cas RAMP module) proteins found in*Pyrococcus furiosus* and other prokaryotes form a functional complex with small CRISPR RNAs that recognizes and cleaves complementary target RNAs.

The Significance for Evolution and Possible Applications

A bioinformatic study has shown that the CRISPRs are evolutionarily conserved and cluster into related types. Many show signs of a conserved secondary structure.

Through the CRISPR-Cas mechanism bacteria can acquire immunity against certain phages and thus halt further transmission of targeted phages. For this reason, some researchers have proposed that the CRISPR-Cas system is a Lamarckian inheritance mechanism. Others investigated the coevolution of host and viral genomes by analysis of CRISPR sequences.

There are several proposals to exploit CRISPRs-derived biotechnology:

- Artificial immunization against phages by adding matching spacers of industrially important bacteria, such as in the milk - and wine industry.
- Knockdown of endogenous genes by transformation with a plasmid which contains a CRISPR area with a spacer, which inhibits a target gene.
- Distinguish different bacterial strains by comparison of the spacer regions (spoligotyping).

Long Non-coding RNA

Long non-coding RNAs (long ncRNAs) are generally considered (somewhat arbitrarily) as non-protein coding transcripts longer than 200nucleotides. This limit is due to practical considerations including the separation of RNAs in common experimental protocols. Additionally, this limit distinguishes long ncRNAs from small regulatory RNAs such as microRNAs (miRNAs), short interfering RNAs (siRNAs), Piwi-interacting RNAs (piRNAs), small nucleolar RNAs (snoRNAs) etc.

Abundance of Long ncRNAs

A recent study found only one fifth of transcription across the human genome is associated with protein-coding genes (Kapranov 2007), indicating at least four-times more long non-coding than coding RNA sequences. However, it is large-scale complementary DNA (cDNA) sequencing projects such as FANTOM (Functional Annotation of Mammalian cDNA) that reveal the complexity of this transcription (Carninci 2005). The FANTOM3 project identified ~35,000 non-coding transcripts from ~10,000 distinct loci that bear many signatures of mRNAs, including 5'capping, splicing and poly-adenylation, but have little or no open reading frame (ORF) (Carninci 2005). While the abundance of long ncRNAs was unanticipated, this number nevertheless represents a conservative lower estimate since it omitted many singleton transcripts and non-polyadenylated transcripts (tiling array data shows more than 40% of transcripts are non-polyadenylated) (Cheng 2005). However, unambiguously identifying ncRNAs within these cDNA libraries is challenging since it can be difficult to distinguish protein-coding transcripts from non-coding transcripts.

Genomic Organisation of Long ncRNAs

The current landscape of the mammalian genome is described as numerous 'foci' of transcription that are separated by long stretches ofintergenic space (Carninci 2005). While long ncRNAs are located and transcribed within the intergenic stretches, the majority are transcribed as complex, interlaced networks of overlapping sense and antisense transcripts that often includes protein-coding genes (Kapranov 2007). Genomic sequences within these transcriptional foci are often shared within a number of different coding and non-coding transcripts in the sense and antisense directions (Birney 2007) giving rise to a complex hierarchy of overlapping isoforms. For example, 3012 out of 8961 cDNAs previously annotated as truncated coding sequences within FANTOM2 were later designated as genuine ncRNA variants of protein-coding cDNAs (Carninci 2005). While the abundance and conservation of these interleaved arrangements suggest they have biological relevance, the complexity of these foci frustrates easy evaluation.

Conservation of Long ncRNAs

Many small RNAs, such as microRNAs or snoRNAs, exhibit strong conservation across diverse species (Bentwich 2005). In contrast, long ncRNAs generally lack strong conservation which is often cited as evidence of non-functionality (Brosius 2005; Struhl 2007). However, many well-described long ncRNAs, such as Air and Xist, are poorly conserved (Nesterova 2001), suggesting that ncRNAs may be subject to different selection pressures (Pang 2006). Unlike mRNAs, which have to conserve the codon usage and prevent frameshift mutations in a single long ORF, selection may conserve only short regions of long ncRNAs that are constrained by structure or sequence-specific interactions. Therefore we may see selection act only over small regions of the long ncRNA transcript. Nevertheless, despite low conservation of long ncRNAs generally, it should be noted that many long ncRNAs still contain strongly conserved elements. For example 19% of highly conserved phastCons elements occur in known introns, and another 32% in unannotated regions (Siepel 2005). Furthermore, a representative set of human long ncRNAs exhibit small, yet significant, reductions in substitution and insertion/deletion rates indicative of purifying selection that conserve the integrity of the transcript at the levels of sequence, promoter and splicing (Ponjavic 2007). The poor conservation of ncRNAs may be the result of recent and rapid adaptive selection. For instance, ncRNAs may be more pliant to evolutionary pressures than protein-coding genes, as evidenced by the existence of many lineage specific ncRNAs, such as Xist or Air (Pang 2006).

Indeed, those conserved regions of the human genome that are subject to recent evolutionary change relative to the chimpanzee genome occurs mainly in non-coding regions, many of which are transcribed. This includes a ncRNA, HAR1F, which has undergone rapid evolutionary change in humans and is specifically expressed in the Cajal-Retzious cells in the human neocortex (Pollard 2006). The observation that many functionally validated RNAs are evolving quickly (Pang 2006; Smith 2004), may result from these sequences having more plastic structure-function constraints, and we may expect a great deal of evolutionary innovation to occur in such sequences. This is supported by the existence of thousands of sequences in the mammalian genome that show poor conservation at the primary sequence level but have evidence of conserved RNA secondary structures.

Long ncRNAs in the Regulation of Gene Transcription

Long ncRNAs in Gene-specific Transcription

In eukaryotes, RNA transcription is a tightly regulated process. NcRNAs can target different aspects of this process, targeting transcriptional activators or repressors, different components of the transcription reaction including RNA polymerase (RNAP) II and even the DNA duplex to regulate gene transcription and expression (Goodrich 2006). In combination these ncRNAs may comprise a regulatory network that, including transcription factors, finely control the gene expression in complex eukaryotes NcRNAs modulate the function of transcription factors by several different mechanisms, including functioning itself as a co-regulator, modifying transcription factor activity, or by regulating the association and activity of co-regulators. For example, the ncRNA Evf-2 functions as a co-activator for the homeobox transcription factor Dlx2, which fulfils important roles in forebrain development and neurogenesis. Sonic hedgehog induces transcription of Evf-2 from an ultraconserved element located between the Dlx5 and Dlx6genes during forebrain development (Feng 2006). Evf-2 then recruits the Dlx2 transcription factor to the same ultraconserved element whereby Dlx2 subsequently induces expression of Dlx5. The existence of other similar ultra- or highly conserved elements within the mammalian genome that are both transcribed and fulfil enhancer function suggest Evf-2 may be illustrative of a generalised mechanism that tightly regulates important developmental genes with complex expression patterns during vertebrate growth. Indeed the transcription and expression of similar non-coding ultraconserved elements was recently shown to be abnormal in human leukaemia and contribute towards apoptosis in colon cancer cells, suggesting an involvement in tumorogenesis (Calin 2007).

Local ncRNAs can also recruit transcriptional programmes to regulate adjacent protein-coding gene expression. The RNA binding protein TLS, binds and inhibits the CREB binding protein and p300 histone acetyltransferease activities on a repressed gene target, cyclin D1. The recruitment of TLS to the promoter of cyclin D1 was directed by lowly expressed long ncRNAs tethered to 5' regulatory regions in response to DNA damage signals (Wang 2008). Moreover, these local ncRNAs act cooperatively as ligands to modulate the activities of TLS. More broadly, this mechanism allows the cell to harness RNA-binding proteins, which make up one of the largest classes within the mammalian proteome, and integrate their function in transcriptional programs.

Recent evidence has raised the possibility that transcription of genes that escape from X-inactivation might be mediated by expression of long non-coding RNA within the escaping chromosomal domains.

Long ncRNAs Regulating Basal Transcription Machinery

NcRNAs also target general transcription factors required for the RNAP II transcription of all genes (Goodrich 2006). These general factors include components of the initiation complex that assemble on promoters or involved in transcription elongation. A ncRNA transcribed from an upstream minor promoter of the dihydrofolate reductase (DHFR) gene forms a stable RNA-DNA triplex within the major promoter of DHFR to prevent the binding of the transcriptional co-factor TFIID (Martianov 2007). This novel mechanism of regulating gene expression may in fact represent a widespread method of controlling promoter usage given that thousands of such triplexes exist in eukaryotic chromosome (Lee 1987).

The U1 ncRNA can induce transcription initiation by specifically binding to and stimulating TFIIH to phosphorylate the C-terminal domain of RNAP II (Kwek 2002). In contrast the ncRNA 7SK, is able to repress transcription elongation by, in combination with HEXIM1/2, forming an inactive complex that prevents the PTEFb general transcription factor from phosphorylating the C-terminal domain of RNAP II, thereby repressing global elongation under stressful conditions. These examples which bypass specific modes of regulation at individual promoters to mediate changes directly at the level of initiation and elongation transcriptional machinery provides a means of quickly affecting global changes in gene expression.

The ability to quickly mediate global changes is also apparent in the rapid expression of non-coding repetitive sequences. The short interspersed nuclear (SINE) Alu elements in humans and analogous B1 and B2 elements in mice have succeeded in becoming the most abundant mobile elements within the genomes, comprising ~10% of the human and ~6% of the mouse genome, respectively. These elements are transcribed as ncRNAs by RNAP III in response to environmental stresses such as heat shock (Liu 1995), where they then bind to RNAP II with high affinity and prevent the formation of active pre-initiation complexes. This allows for the broad and rapid repression of gene expression in response to stress.

A dissection of the functional sequences within Alu RNA transcripts has drafted a modular structure analogous to the organization of domains in protein transcription factors (Shamovsky 2008). The Alu RNA contains two 'arms', each of which may bind one RNAP II molecule, as well as two regulatory domains that are responsible for RNAP II transcriptional repression in vitro (Mariner 2008). These two loosely-structured domains may even be concatenated to other ncRNAs such as B1 elements to impart their repressive role.

The abundance and distribution of Alu elements and similar repetitive elements throughout the mammalian genome may be partly due to these functional domains being co-opted into other long ncRNAs during evolution, with the presence of functional repeat sequence domains being a common characteristic of several known long ncRNAs including Kcnq1ot1, Xlsirt and Xist.

In addition to heat shock, the expression of SINE elements (including Alu, B1 and B2 RNAs) increases during cellular stress such as viral infection (Singh 1985) in some cancer cells (Tang 2005) where they may similarly regulate global changes to gene expression.

The ability of Alu and B2 RNA to bind directly to RNAP II provides a broad mechanism for to repress transcription. Nevertheless, there are specific exceptions to this global response where Alu or B2 RNAs are not found at activated promoters of genes undergoing induction, such as the heat shock genes.

This additional hierarchy of regulation that exempts individual genes from the generalised repression also involves a long ncRNA, heat shock RNA-1 (HSR-1). HSR-1 is present in all cells in an inactive state, but upon stress is activated to induce the expression of heat shock genes (Shamovsky 2006). This activation involves a conformational alteration to the structure of HSR-1 in response to rising temperatures, thereby permitting its interaction with the transcriptional activator HSF-1 that subsequently undergoes trimerisation and induces the expression of heat shock genes (Shamovsky 2006).

More broadly, these examples illustrate a regulatory circuit nested witin ncRNAs whereby Alu or B2 RNAs repress general gene expression, while other ncRNAs, such as HSR-1, activate the expression of specific genes.

Long ncRNA Transcribed by RNA Polymerase III

Many of the ncRNAs that interact with general transcription factors or RNAP II itself (including 7SK, Alu and B1 and B2 RNAs) are transcribed by RNAP III (Dieci 2007), thereby decoupling the expression of these ncRNAs from the RNAP II transcriptional reaction they regulate. RNAP III also transcribes a number of additional novel ncRNAs, such as BC2, BC200 and some microRNAs and snoRNAs, in addition to the highly-expressed infrastructural 'housekeeping' ncRNA genes such as tRNAs, 5S rRNAs and snRNAs (Dieci 2007).

The existence of an RNAP III-dependent ncRNA transcriptome that regulates its RNAP II-dependent counterpart was supported by a recent study that described a novel set of ncRNAs transcribed by RNAP III with sequence homology to protein-coding genes. This prompted the authors to posit a 'cogene/gene' functional regulatory network (Pagano 2007), showing that one of these ncRNAs, 21A, regulates the expression its antisense partner gene, CENP-F in trans.

Long Non-coding RNAs in Post-transcriptional Regulation

In addition to regulating transcription, ncRNAs also control various aspects of post-transcriptional mRNA processing. Similar to small regulatory RNAs such as microRNAs and snoRNAs, these functions often involve complementary base pairing with the target mRNA. The formation of RNA duplexes between complementary ncRNA and mRNA may mask key elements within the mRNA required to bind trans-acting factors, potentially effecting any step in post-transcriptional gene expression including pre-mRNA processing and splicing, transport, translation, and degradation.

Long ncRNAs in Splicing

The splicing of mRNA can induce its translation and functionally diversify the repertoire of proteins it encodes. The Zeb2 mRNA, which has a particularly long 5'UTR, requires the retention of a 5'UTR intron that contains an internal ribosome entry site for efficient translation (Beltran 2008). However, retention of the intron is dependent on the expression of an antisense transcript that complements the intronic 5' splice site (Beltran 2008). Therefore, the ectopic expression of the antisense transcript represses splicing and induces translation of the Zeb2 mRNA during mesenchymal development. Similarly, the expression of an overlapping antisense Rev-ErbAa2 transcript controls the alternative splicing of the thyroid hormone receptor ErbAa2 mRNA to form two antagonistic isoforms (Munroe 1991).

Long ncRNAs in Translation

NcRNA may also apply additional regulatory pressures during translation, a property particularly exploited in neurons where the dendritic or axonal translation of mRNA in response to synaptic activity contributes to changes in synaptic plasticity and the remodelling of neuronal networks. The RNAP III transcribed BC1 and BC200 ncRNAs, that previously derived from tRNAs, are expressed in the mouse and human central nervous system, respectively. BC1 expression is induced in response to synaptic activity and synaptogenesis and is specifically targeted to dendrites in neurons (Muslimov 1998). Sequence complementarity between BC1 and regions of various neuron-specific mRNAs also suggest a role for BC1 in targeted translational repression (Wang 2005). Indeed it was recently shown that BC1 is associated with translational repression in dendrites to control the efficiency of dopamine D2 receptor-mediated transmission in the striatum (Centonze 2007) and BC1 RNA-deleted mice exhibit behavioural changes with reduced exploration and increased anxiety (Lewejohann 2004).

Long ncRNAs in siRNA-directed Gene Regulation

In addition to masking key elements within single-stranded RNA, the formation of double stranded RNA duplexes can also provide a substrate for the generation of endogenous siRNAs (endo-siRNAs) in Drosophila and mouse oocytes (Golden 2008). The annealing of complementary sequences, such as antisense or repetitive regions between transcripts, forms a RNA duplex that may be processing by for Dicer-2 into endo-siRNAs. Alternatively, long ncRNAs that form extended intramolecular hairpins may also be processed into siRNAs, compellingly illustrated by the esi-1 and esi-2 transcripts (Czech 2008). Endo-siRNAs generated from these transcripts seem particularly useful in suppressing the spread of mobile transposon elements within the genome in the germline. However, the generation of endo-siRNAs from antisense transcripts or pseudogenes may also silence the expression of their functional counterparts via RISC effector complexes, acting as an important node that integrates integrate various modes of long and short RNA regulation, as exemplified by the Xist and Tsix (Ogawa 2008).

Long ncRNAs in Epigenetic Regulation

Epigenetic modifications, including histone and DNA methylation, histone acetylation and sumoylation, affect many aspects of chromosomal biology, primarily including regulation of large numbers of genes by remodeling broad chromatin domains. While it has been known for some time that RNA is an integral component of chromatin, it is only recently that we are beginning to appreciate the means by which RNA is involved in pathways of chromatin modification.

In Drosophila, long ncRNAs induce the expression of the homeotic gene, Ubx, by recruiting and directing the chromatin modifying functions of the trithorax protein Ash1 to Hox regulatory elements (Sanchez-Elsner 2006).

Similar models have been proposed in mammals, where strong epigenetic mechanisms are thought to underlie the embryonic expression profiles of the Hox genes that persist throughout human development. Indeed, the human Hox genes are associated with hundreds of ncRNAs that are sequentially expressed along both the spatial and temporal axes of human development and define chromatin domains of differential histone methylation and RNA polymerase accessibility (Rinn 2007). One ncRNA, termed HOTAIR, that originates from the HOXC locus represses transcription across 40 kb of the HOXD locus by altering chromatin trimethylation state. HOTAIR is thought to achieve this by directing the action of Polycomb chromatin remodeling complexes in trans to govern the cells epigenetic state and subsequent gene expression. Components of the Polycomb complex, including Suz12, EZH2 and EED, contain RNA binding domains that may potentially bind HOTAIR and probably other similar ncRNAs. This example nicely illustrates a broader theme whereby ncRNAs recruit the function of a generic suite of chromatin modifying proteins to specific genomic loci, underscoring the complexity of recently published genomic maps (Mikkelsen 2007). Indeed the prevalence of long ncRNAs associated with protein coding genes may contribute to localised patterns of chromatin modifications that regulate gene expression during development. For example, the majority of protein-coding genes have antisense partners, including many tumour suppressor genes that are frequently silenced by epigenetic mechanisms in cancer (Yu 2008). A recent study observed an inverse expression profile of the p15 gene and an antisense ncRNA in leukaemia (Yu 2008). A detailed analysis showed the p15 antisense ncRNA was able to induce changes to heterochromatin and DNA methylation status of p15 by an unknown mechanism, thereby regulating p15 expression (Yu 2008). Therefore misexpression of the associated antisense ncRNAs may subsequently silence the tumour suppressor gene contributing towards oncogenesis.

Imprinting

Many emergent themes of ncRNA-directed chromatin modification were first apparent within the phenomenon of imprinting, whereby only one allele of a gene is expressed from either the maternal or paternal chromosome. Imprinted genes are generally clustered together on chromosomes, suggesting the imprinting mechanism acts upon local chromosome domains rather than individual genes. These clusters are also often associated with long ncRNAs whose expression is correlated with the repression of the linked protein-coding gene on the same allele (Pauler 2007). Indeed, detailed analysis has revealed a crucial role for the ncRNAs Kcnqot1 and Igf2r/Air in directing imprinting (Braidotti 2004).

Almost all the genes at the Kcnq1 loci are maternally inherited, except the paternally expressed antisense ncRNA Kcnqot1 (Mitsuya 1999). Transgenic mice with truncated Kcnq1ot fail to silence the adjacent genes, suggesting that Kcnqot1 is crucial to the imprinting of genes on the paternal chromosome. It appears that Kcnqot1 is able to direct the trimethylation of lysine 9 (H3K9me3) and 27 of histone 3 (H3K27me3) to an imprinting centre that overlaps the Kcnqot1 promoter and actually resides within a Kcnq1 sense exon (Umlauf 2004). Similar to HOTAIR, Eed-Ezh2 Polycomb complexes are recruited to the Kcnq1 loci paternal chromosome, possibly by Kcnqot1, where they may mediate gene silencing through repressive histone methylation (Umlauf 2004). A differentially methylated imprinting centre also overlaps the promoter of a long antisense ncRNA Air that is responsible for the silencing of neighbouring genes at the Igf2r locus on the paternal chromosome. The presence of allele-specific histone methylation at the Igf2r locus suggests Air also mediates silencing via chromatin modification (Fournier 2002).

Xist and X-chromosome Inactivation

The inactivation of a X-chromosome in female placental mammals is directed by one of the earliest and best characterized long ncRNAs, Xist(Wutz 2007). The expression of Xist from the future inactive X-chromosome, and its subsequent coating of the inactive X-chromosome, occurs during early embryonic stem cell differentiation. Xist expression is followed by irreversible layers of chromatin modifications that include the loss of the histone (H3K9) acetylation and H3K4 methylation that are associated with active chromatin, and the induction of repressive chromatin modifications including H4 hypoacetylation, H3K27 trimethylation (Wutz 2007), H3K9 hypermethylation and H4K20 monomethylation as well as H2AK119 monoubiquitylation. These modifications coincide with the transcriptional silencing of the X-linked genes (Morey 2004). Xist RNA also localises the histone variant macroH2A to the inactive X–chromosome (Costanzi 1998). There are additional ncRNAs that are also present at the Xist loci, including an antisense transcript Tsix, which is expressed from the future active chromosome and able to repress Xist expression by the generation of endogenous siRNA (Ogawa 2008). Together these ncRNAs ensure that only one X-chromosome is active in female mammals.

Telomeric Non-coding RNAs

Telomeres form the terminal region of mammalian chromosomes and are essential for stability and aging and play central roles in diseases such as cancer (Blasco 2007). Telomeres have been long considered transcriptionally inert DNA-protein complexes until it was recently shown that telomeric repeats may be transcribed as telomeric RNAs (TelRNAs) (Schoeftner 2008) or telomeric repeat-containing RNAs (Azzalin 2007). These ncRNAs are heterogeneous in length, transcribed from several sub-telomeric loci and physically localise to telomeres. Their association with chromatin, which suggests an involvement in regulating telomere specific heterochromatin modifications, is repressed by SMG proteins that protect chromosome ends from telomere loss (Azzalin 2007). In addition, TelRNAs block telomerase activity in vitro and may therefore regulate telomerase activity (Schoeftner 2008). Although early, these studies suggest an involvement for telomeric ncRNAs in various aspects of telomere biology.

Long Non-coding RNAs in Disease

Recent recognition that long ncRNAs function in various aspects of cell biology has focused increasing attention on their potential to contribute towards disease aetiology. A handful of studies have implicated long ncRNAs in a variety of disease states and support an involvement and co-operation in oncogenesis.

While many association studies have identified long ncRNAs that are aberrantly expressed in disease states, we have little understanding of their contribution within disease etiology. Expression analyses that compare tumour cells and normal cells have revealed changes in the expression of ncRNAs in several forms of cancer. For example, the ncRNA OCC-1 (overexpressed in colon carcinoma-1) is overexpressed in colon carcinoma cells (Pibouin 2002). Similarly, in prostate tumours, one of two overexpressed ncRNAs, PCGEM1, is correlated with increased proliferation and colony formation suggesting an involvement in regulating cell growth (Fu 2006). NEAT2 was originally identified as an abundantly expressed ncRNA that is upregulated during metastasis of early-stage non-small cell lung cancer and its overexpression is an early prognostic marker for poor patient survival rates (Fu 2006). More recently, the highly conserved mouse homologue of NEAT2 was found to be highly expressed in hepatocellular carcinoma (Lin 2007). Intronic antisense ncRNAs with expression correlated to the degree of tumour differentiation in prostate cancer samples have also been reported (Reis 2004). Despite a number of long ncRNAs having aberrant expression in cancer, their function and potential role in tumourogenesis is relatively unknown. For example, the ncRNAs HIS-1 and BIC have been implicated in oncogenesis and growth control, but their function in normal cells is unknown (Eis 2005; Li 1997). In addition to cancer, ncRNAs also exhibit aberrant expression in other disease states. Overexpression of PRINS is associated with psoriasis susceptibility, with PRINS expression being elevated in the uninvolved epidermis of psoriatic patients compared with both psoriatic lesions and healthy epidermis (Sonkoly 2005).

Genome-wide profiling revealed that many transcribed non-coding ultraconserved regions exhibit distinct profiles in various human cancer states (Calin 2007). An analysis of chronic lymphocytic leukaemia, colorectal carcinoma and hepatocellular carcinoma found that all three cancers exhibited aberrant expression profiles for ultraconserved ncRNAs relative to normal cells. Further analysis of one ultraconserved ncRNA suggested it behaved like an oncogene by mitigating apoptosis and subsequently expanding the number of malignant cells in colorectal cancers (Calin 2007). Many of these transcribed ultraconserved sites that exhibit distinct signatures in cancer are found at fragile sites and genomic regions associated with cancer. It seems likely that the aberrant expression of these ultraconserved ncRNAs within malignant processes results from important functions they fulfil in normal human development.

Recently a number of association studies examining single nucleotide polymorphisms (SNPs) associated with disease states have been mapped to long ncRNAs. For example, SNPs that identified a susceptibility locus for myocardial infarction mapped to a long ncRNA, MIAT (myocardial infarction associated transcript) (Ishii 2006). Similarly, genome-wide association studies identified a region associated with coronary artery disease (McPherson 2007) that encompassed a long ncRNA, ANRIL (Pasmant 2007). ANRIL is expressed in tissues and cell types affected by atherosclerosis and its altered expression is associated with a high-risk haplotype for coronary artery disease.

The complexity of the transcriptome, and our evolving understanding of its structure may inform a reinterpretation of the functional basis for many natural polymorphisms associated with disease states. Many SNPs associated with certain disease conditions are found within non-coding regions and the complex networks of non-coding transcription within these regions make it particularly difficult to elucidate the functional effects of polymorphisms. For example, a SNP both within the truncated form of ZFAT and the promoter of an antisense transcript increases the expression of ZFAT not through increasing the mRNA stability, but rather by repressing the expression of the antisense transcript (Shirasawa 2004).

The ability of long ncRNAs in regulating associated protein-coding genes may contribute towards disease if the long ncRNAs misexpression deregulates a protein coding genes with clinical significance. Similarly, an antisense long ncRNA that regulates the expression of the sense BACE1 gene, a crucial enzyme in Alzheimer's disease etiology, exhibits elevated expression in several regions of the brain in individuals with Alzheimer's disease (Faghihi 2008). Alteration of the expression of ncRNAs may also mediate changes at an epigenetic level to effect gene expression and contribute to disease aetiology. For example, the induction of an antisense transcript by a genetic mutation led to the DNA methylation and silencing of sense genes, causing β-thalassemia in a patient (Tufarelli 2003).

Piwi-interacting RNA

Piwi-interacting RNA (piRNA) is the largest class of small RNA molecules that is expressed in animal cells. piRNA forms RNA-protein complexes through interactions with Piwi proteins. These piRNA complexes have been linked to transcriptional gene silencing ofretrotransposons and other genetic elements in germ line cells, particularly those in spermatogenesis. They are distinct from miRNA in size (26–31 nt rather than 21–24 nt), lack of sequence conservation, and increased complexity.

It remains unclear how piRNAs are generated, but potential methods have been suggested, and it is certain their biogenesis pathway is distinct from miRNA and siRNA, while rasiRNAs are a piRNA subspecies.

Characteristics

piRNAs have been identified in both vertebratesand invertebrates, and although biogenesis and modes of action do vary somewhat betweenspecies, a number of features are conserved. piRNAs have no clear secondary structuremotifs,. the length of a piRNA is, by definition, between 26 and 31 nucleotides, and the presence of a 5' uridine is common to piRNAs in both vertebrates and invertebrates. piRNAs in *C. elegans* have a 5' monophosphate and a 3' modification that acts to block either the 2' or 3' oxygen, and this has also been confirmed to exist in *D.melanogaster*, zebrafish, mice and rats.

This 3' modification is likely to be a 2'-O-methylation, but the reason for this modification is not known. It is thought that there are many hundreds of thousands of different piRNA species found inmammals. Thus far, over 50,000 unique piRNA sequences have been discovered in mice and more than 13,000 in *D.melanogaster*.

Location

piRNAs are found in clusters throughout the genome; these clusters may contain as few as ten or up to many thousands of piRNAs and can vary in size from one to one hundred kb. While the clustering of piRNAs is highly conserved across species, the sequences are not. While *D.melanogaster* and vertebrate piRNAs have been located in areas lacking any protein coding genes, piRNAs in *C. elegans* have been identified amidst protein coding genes.

In mammals, piRNAs are found only within the testes, with an estimated one million copies per cell in spermatocytes and spermatids. In invertebrates, piRNAs have been detected in both the male and female germlines, but in no other cell types.

At the cellular level, piRNAs have been found within both nuclei and cytoplasm, suggesting that piRNA pathways may function in both of these areas and, therefore, may have multiple effects.

Biogenesis

The biogenesis of piRNAs is not yet fully understood, although possible mechanisms have been proposed. piRNAs show a significant strand bias, that is, they are derived from one strand of DNA only, and this may indicate that they are the product of long single stranded precursor molecules. A primary processing pathway is suggested to be the only pathway used to produce pachytene piRNAs; in this mechanism, piRNA precusors are transcribed resulting in piRNAs with a tendency to target 5' uridines.

Also proposed is a 'Ping Pong' mechanism wherein primary piRNAs recognise their complementary targets and cause the recruitment of Piwi proteins. This results in the cleavage of the transcript at a point ten nucleotides from the 5' end of the primary piRNA, producing the secondary piRNA.

These secondary piRNAs are targeted toward sequences that possess an adenine at the tenth position. Since the piRNA involved in the ping pong cycle directs its attacks on transposon transcripts, the ping pong cycle acts only at the level of transcription. One or both of these mechanisms may be acting in differentspecies; *C. elegans*, for instance, does have piRNAs, but does not appear to use the ping pong mechanism at all.

A significant number of piRNAs identified in zebrafish and *D.melanogaster* contain adenine at their tenth position, and this has been interpreted as possible evidence of a conserved biosynthetic mechanism across species. piRNAs are expressed through unique pathways, which are dissimilar to the expression pathways utilised by other small RNAs. Neither Argonaute proteins, RNA-dependent RNA polymerases, nor other proteins required for most small-RNA biosynthesis are necessary for piRNA expression. However, while PIWI proteins are not required for the biosynthesis of piRNAs, evidence suggests that PIWI proteins must be present to stabilise piRNAs and, thus, facilitate their accumulation. No mechanism for the control of piRNA propagation has yet been established.

Function

The wide variation in piRNA sequences and PIWI function over species contributes to the difficulty in establishing the functionality of piRNAs. However, like other small RNAs, piRNAs are thought to be involved in gene silencing, specifically the silencing of transposons. The majority of piRNAs are antisense to transposon sequences, suggesting that transposons are the piRNA target. In mammals it appears that the activity of piRNAs in transposon silencing is most important during the development of the embryo, and in both *C. elegans* and humans, piRNAs are necessary for spermatogenesis.

RNA Silencing

piRNA has a role in RNA silencing via the formation of an RNA-induced silencing complex (RISC). piRNAs interact with Piwi proteins that are part of a family of proteins called the Argonautes. These are active in the testes of mammals and are required for germ-cell and stem-celldevelopment in invertebrates. Three Piwi subfamily proteins - MIWI, MIWI2 and MILI - have been found to be essential for spermatogenesis inmice. piRNAs direct the Piwi proteins to their transposon targets. A decrease or absence of PIWI protein expression is correlated with an increased expression of transposons.

Transposons have a high potential to cause deleterious effect on their host, and, in fact, mutations in piRNA pathways are found to reduce fertility in *D.melanogaster*. However, piRNA pathway mutations in mice do not demonstrate reduced fertility; this may indicate redundancies to the piRNA system. Further, it is thought that piRNA and endogenous small interfering RNA (endo-siRNA) may have comparable and even redundant functionality in transposon control in mammalian oocytes.

piRNAs appear to have an impact on particular methyltransferases that perform the methylations which are required to recognise and silence transposons, but this relationship is not well understood.

5

Trans-acting RNA

Trans-acting siRNA (ta-siRNA or tasiRNA) is a form of small interfering RNA (siRNA) that represses gene expression through post-transcriptional gene silencing in land plants. It is transcribed from the genome to form a polyadenylated, double-stranded precursor. Unlike other siRNAs, but like microRNA, ta-siRNA does not require full complementarity to its target RNA to have an effect.

RasiRNA

Repeat associated small interfering RNA (rasiRNA) is a class of small RNA that is involved in the RNA interference (RNAi) pathway. RasiRNA is a subclass of Piwi-interacting RNAs (piRNAs), which are small RNA molecule which interact with Piwi proteins. Piwi proteins are an Argonaute subfamily. RasiRNA associate with both the Ago and Piwi Argonaute protein subfamily while piRNA only associates with the Piwi Argonaute subfamily. In the germline, RasiRNA is involved in establishing and maintaining heterochromatin structure, controlling transcripts that emerge from repeat sequences, and silencing transposons and retrotransposons.

Classification

There are at least three Argonaute subfamilies that have been found in eukaryotes. Unlike the Ago subfamily which is present in animals, plants, and fission yeast, the Piwi subfamily has only been found in animals. RasiRNA has been observed in *Drosophila* and some unicellular eukaryotes but its presence in mammals has not been determined unlike piRNA which has been observed in many species of invertebrates and vertebrates including mammals;

however, since proteins which associate with rasiRNA are found in both vertebrates and invertebrates, it is possible that active rasiRNA exist and have yet to be observed in other animals. RasiRNAs have been observed inSchizosaccharomyces pombe, a species of yeast, as well in some plants both of which have not been observed to contain the Piwi Argonaute protein subfamily. It has been observed that both rasiRNA and piRNA are maternally linked, but more specifically it is the Piwi protein subfamily that is maternally linked and therefore leads to the observation that rasiRNA and piRNA are maternally linked.

RasiRNA are Distinct from other RNAi Pathways

RasiRNA is distinct from other RNAi pathways such as microRNA (miRNA) and small interfering RNA (siRNA) as well as from piRNA. Unlike miRNA and siRNA which function through the Ago Argonaute protein subfamily, RasiRNA function through the Piwi Argonaute protein subfamily. RasiRNA is also distinct in its size. Contrary to miRNAs which are 21-23 nucleotides in length, siRNAs which are 20-25 nucleotides in length, and piRNAs which are 24-31 nucleotides in length, rasiRNAs are 24-29 nucleotides in length depending on the organism it came from. Unlike siRNA which are derived from both the sense and antisense strand, rasiRNA are derived from the antisense. Interestingly, while miRNA requires Dicer-1 for its production, and siRNA requires Dicer-2, rasiRNA does not require either; however, in some plants there are Dicer-like (Dcl) proteins that have been identified where Dcl1 produces 24 nucleotide miRNA and siRNA while Dcl2 produces 24 nucleotide rasiRNA. This research shows that not only is rasiRNA production distinct from miRNA and siRNA, but that rasiRNA may be found in plants while Piwi proteins are not.

RasiRNA Biogenesis

It is presumed that the source of rasiRNA is double stranded RNA produced by annealing of sense and antisense related transposable elements. The biogenesis of rasiRNA is independent of Dicer, but does require the Argonaute proteins Argonaute 3 (Ago 3), Piwi and Aubergine which is a Piwi-like protein. The mechanism for rasiRNA biogenesis is a ping-pong mechanism. The Piwi/ Aub associated RNA is the rasiRNA. The rasiRNAs match the antisense strand of retrotransposons and repetitive sequence elements (hence the name rasiRNA). The Ago3 associated RNAs are derived from the sense strand. The ping-pong mechanism which is observed in this image is the mechanism for the generation of the 5' end of rasiRNA while the generation of the 3' end of rasiRNA is still unknown.

Importance of RasiRNA

While miRNA act in translational repression and mRNA cleavage and siRNA act in mRNA cleavage, rasiRNA act to regulate chromatin structure and transcriptional silencing. In Drosophila, mutations in the Piwi proteins that associate with rasiRNA lead to sterility and loss of germline cells in both males and females. Transposon repression is not affected by the loss of Dicer within the germline cells revealing that this is the target of the rasiRNA pathway. Similar to miRNA and siRNA, the rasiRNA silencing pathway is evolutionarily conserved and homology dependent. When the rasiRNA pathway is not present, germline cells may undergo retrotransposition which are sensed as DNA damage and signal the cell to apoptosis. RasiRNA is key to the regulatory mechanism of many organisms as part of the RNA interference pathway.

History and Discovery

Small RNAs guiding RNA silencing pathways were first discovered in 1993 in Caenorhabditis elegans and since then have been observed in many organisms. RasiRNA were discovered sometime after 1993 and observed in both Drosophila melanogaster andSchizosaccharomyces pombe and since have been observed in some single celled eukaryotes and some plants.

7SK RNA

7SK, also known as RN7SK, was discovered as an abundant small nuclear RNA gene found in metazoans in the mid-70s but a possible function has only recently been suggested.

Two independent studies have found that 7SK RNA binds the CDK9/cyclin T complex (known as elongation factor P-TEFb). P-TEFb activates transcription byphosphorylating the C-terminal domain of RNA polymerase II. This process is negatively regulated by the 7SK RNP.

The secondary structure of 7SK RNA shown here was proposed by Wasserman and Steitz on the basis of chemical modification and enzymatic cleavage experiments. This family contains a number of pseudogenes and 7SK derived repeats.

Retrotransposon

Retrotransposons (also called transposons via RNA intermediates) are genetic elements that can amplify themselves in a genome and are ubiquitous components of the DNA of many eukaryotic organisms. They are a subclass of transposon. They are particularly abundant in plants, where they are often a principal component of nuclear DNA. In maize, 49-78% of the genome is made up of retrotransposons. In wheat, about 90% of the genome consists of repeated sequences and 68% of transposable elements. In mammals, almost half the genome (45% to 48%) comprises transposons or remnants of transposons. Around 42% of the human genome is made up of retrotransposons while DNA transposons account for about 2-3%.

Biological Activity

The retrotransposons' replicative mode of transposition through an RNA intermediate increases the copy numbers of elements rapidly and thereby can increase genome size. Like DNA transposable elements (class II transposons), retrotransposons can induce mutations byinserting near or within genes. Furthermore, retrotransposon-induced mutations are relatively stable, because the sequence at the insertion site is retained as they transpose via the replication mechanism.

Retrotransposons copy themselves to RNA and then back to DNA that may integrate back to the genome. The second step of forming DNA may be carried out by a reverse transcriptase which the retrotransposon encodes. Transposition and survival of retrotransposons within the host genome are possibly regulated both by retrotransposon- and host-encoded factors, to avoid deleterious effects on host and retrotransposon as well, in a relationship that has existed for many millions of years between retrotransposons and their plant hosts. The understanding of how retrotransposons and their hosts' genomes have co-evolved mechanisms to regulate transposition, insertion specificities, and mutational outcomes in order to optimize each other's survival is still in its infancy.

Most retrotransposons are very old and through accumulated mutations, are no longer able to retrotranspose.

Types of Retrotransposons

Retrotransposons, also known as class I transposable elements, consist of two sub-types, the long terminal repeat (LTR) and the non-LTR retrotransposons.

LTR Retrotransposons

LTR retrotransposons have direct LTRs that range from ~100 bp to over 5 kb in size. LTR retrotransposons are further sub-classified into the Ty1-copia-like (Pseudoviridae), Ty3-gypsy-like (Metaviridae), and Pao-BEL-like groups based on both their degree of sequence similarity and the order of encoded gene products. Ty1-copia and Ty3-gypsy groups of retrotransposons are commonly found in high copy number (up to a few million copies per haploid nucleus) in animals, fungi, protista, and plants genomes. Pao-BEL like elements have so far only been found in animals. About 8% of the human genome and approximately 10% of the mouse genome are composed of the LTR transposons.

Ty1-copia Retrotransposons

Are abundant in species ranging from single-cell algae to bryophytes, gymnosperms, and angiosperms.

Ty3-gypsy Retrotransposons

Are also widely distributed, including both gymnosperms and angiosperms.

Non-LTR Retrotransposons

Consists of two sub-types, long interspersed nuclear elements (LINEs) and short interspersed nuclear elements (SINEs). They can also be found in high copy numbers (up to 250,000) in the plant species.

LINEs

Long interspersed repetitive elements or Long interspersed nuclear elements are a group of genetic elements that are found in large numbers in eukaryotic genomes. They are transcribed (or are the evolutionary remains of what was once transcribed) to an RNA using anRNA polymerase II promoter that resides inside the LINE. LINEs code for the enzyme reverse transcriptase, and many LINEs also code for an endonuclease (e.g. RNase H). The reverse transcriptase has a higher specificity for the LINE RNA than other RNA, and makes a DNA copy of the RNA that can be integrated into the genome at a new site.

The 5' UTR contains the promoter sequence, while the 3' UTR contains a polyadenylation signal (AATAAA) and a poly-A tail. Because LINEs move by copying themselves (instead of moving, like transposons do), they enlarge the genome. The human genome, for example, contains about 500,000 LINEs, which is roughly 17% of the genome. Of these, approximately 7,000 are full-length, a small subset of which are capable of retrotransposition.

SINEs

Short interspersed repetitive elements or Short interspersed nuclear elements are short DNA sequences (<500 bases) that represent reverse-transcribed RNA molecules originally transcribed by RNA polymerase III into tRNA, rRNA, and other small nuclear RNAs. SINEs do not encode a functional reverse transcriptase protein and rely on other mobile elements for transposition. The most common SINEs in primates are called Alu sequences. Alu elements are 280 base pairs long, do not contain any coding sequences, and can be recognized by the restriction enzyme AluI (hence the name). With about 1,500,000 copies, SINEs make up about 11% of the human genome. While historically viewed as "junk DNA", recent research suggests that in some rare cases both LINEs and SINEs were incorporated into novel genes, so as to evolve new functionality.. The distribution of these elements has been implicated in some genetic diseases and cancers.

Composite SINE Transposons

Two SINES may act in concert to flank and mobilize an intervening single copy DNA sequence. This was reported for a 710 bp DNA sequence upstream of the bovine beta globin gene. The DNA arrangement forms a composite transposon whose presence has been confirmed by the complete bovine genomic sequence where the mobilized sequence may be found on bovine chromosome 15 in contigNW_001493315.1 nucleotides #1085432–1086142 and the originating sequence may be found on bovine chromosome 2 in contigNW_001501789.2 nucleotides #1096679–1097389. It is likely that similar composite transposons exist in other bovine genomic regions and other mammalian genomes. They could be detected with suitable algorithms.

Viroid

Viroids are plant pathogens that consist of a short stretch (a few hundred nucleobases) of highly complementary, circular, single-stranded RNA without the protein coat that is typical for viruses. The smallest discovered is a 220 nucleobase scRNA (small cytoplasmic RNA) associated with the rice yellow mottle sobemovirus (RYMV). In comparison, the genome of the smallest known viruses capable of causing an infection by themselves are around 2 kilobases in size. The human pathogen hepatitis D is similar to viroids.

Viroids were discovered and given this name by Theodor Otto Diener, a plant pathologist at the Agricultural Research Service in Maryland, in 1971. Viroid RNA does not code for any protein. The replication mechanism involves RNA polymerase II, an enzyme normally associated with synthesis of messenger RNA from DNA, which instead catalyses "rolling circle" synthesis of new RNA using the viroid's RNA as template. Some viroids are ribozymes, having catalytic properties which allow self-cleavage and ligation of unit-size genomes from larger replication intermediates.

The first viroid to be identified was the *potato spindle tuber viroid* (PSTVd). Some 33 species have been identified.

Primary Structure

1	CGGAACUAAA	CUCGUGGUUC	CUGUGGUUCA	CACCUGACCU
	CCUGAGCAGA	AAAGAAAAAA		
61	GAAGGCGGCU	CGGAGGAGCG	CUUCAGGGAU	CCCCGGGGAA
	ACCUGGAGCG	AACUGGCAAA		
121	AAAGGACGGU	GGGGAGUGCC	CAGCGGCCGA	CAGGAGUAAU
	UCCCGCCGAA	ACAGGGUUUU		
181	CACCCUUCCU	UUCUUCGGGU	GUCCUUCCUC	GCGCCCGCAG
	GACCACCCCU	CGCCCCCUUU		
241	GCGCUGUCGC	UUCGGCUACU	ACCCGGUGGA	AACAACUGAA
	GCUCCCGAGA	ACCGCUUUUU		
301	CUCUAUCUUA	CUUGCUUCGG	GGCGAGGGUG	UUUAGCCCUU
	GGAACCGCAG	UUGGUUCCU		

Secondary Structure

Taxonomy

- Family Pospiviroidae:
  - o Genus *Pospiviroid*; type species: *Potato spindle tuber viroid*
  - o Genus *Hostuviroid*; type species: *Hop stunt viroid*
  - o Genus *Cocadviroid*; type species: *Coconut cadang-cadang viroid*
  - o Genus *Apscaviroid*; type species: *Apple scar skin viroid*
  - o Genus *Coleviroid*; type species: *Coleus blumei viroid 1.*
- Family Avsunviroidae;
  - o Genus *Avsunviroid*; type species: *Avocado sunblotch viroid*
  - o Genus *Pelamoviroid*; type species: *Peach latent mosaic viroid.*

Viroids and RNA Silencing

There has long been confusion over how viroids are able to induce symptoms in plants without encoding any protein products within their sequences. Evidence now suggests that RNA silencing is involved in the process. First, changes to the viroid genome can dramatically alter its virulence.

This reflects the fact that any siRNAs produced would have less complementary base pairing with target messenger RNA. Secondly, siRNAs corresponding to sequences from viroid genomes have been isolated from infected plants.

Finally, transgenicexpression of the noninfectious hpRNA of potato spindle tuber viroid develops all the corresponding viroid like symptoms.

This evidence indicates that when viroids replicate via a double stranded intermediate RNA, they are targeted by a dicer enzyme and cleaved into siRNAs that are then loaded onto the RNA-induced silencing complex.

The viroid siRNAs actually contain sequences capable of complementary base pairing with the plant's own messenger RNAs and induction of degradation or inhibition of translation is what causes the classic viroid symptoms.

Satellite (Biology)

A satellite is a subviral agent composed of nucleic acid that depends on the co-infection of a host cell with a helper or master virus for its multiplication. When a satellite encodes the coat protein in which its nucleic acid is encapsidated it is referred to as a satellite virus.

A satellite virus of mimivirus that inhibits the replication of its host has been termed a virophage. Satellite viral particles should not be confused with satellite DNA.

Classification

Satellite viruses:

- Single-stranded RNA satellite viruses
- Subgroup 1: Chronic bee-paralysis virus associated satellite
- *Chronic bee-paralysis satellite virus*
- Subgroup 2: Tobacco necrosis virus satellite
- *Maize white line mosaic satellite virus*
- *Panicum mosaic satellite virus*
- *Tobacco mosaic satellite virus*
- *Tobacco necrosis satellite virus.*

Satellite nucleic acids;

- Single-stranded satellite DNAs
- *Tomato leaf curl virus satellite DNA*
- Double-stranded satellite RNAs
- *Saccharomyces cerevisiae M virus satellite*
- *Trichomonas vaginalis T1 virus satellite*
- Single-stranded satellite RNAs
- Subgroup 1: Large satellite RNAs
- *Arabis mosaic virus large satellite RNA*
- *Bamboo mosaic virus satellite RNA*
- *Chicory yellow mottle virus large satellite RNA*
- *Grapevine Bulgarian latent virus satellite RNA*
- *Grapevine fanleaf virus satellite RNA*
- *Myrobalan latent ringspot virus satellite RNA*
- *Tomato black ring virus satellite RNA*
- *Beet ringspot virus satellite RNA*
- Subgroup 2: Small linear satellite RNAs
- *Cucumber mosaic virus satellite RNA*
- *Cymbidium ringspot virus satellite RNA*
- *Pea enation mosaic virus satellite RNA*
- *Groundnut rosette virus satellite RNA*
- *Panicum mosaic virus small satellite RNA*
- *Peanut stunt virus satellite RNA*

- *Turnip crinkle virus satellite RNA*
- *Tomato bushy stunt virus satellite RNA, B10*
- *Tomato bushy stunt virus satellite RNA, B1*
- Subgroup 3: Circular satellite RNAs, or "virusoids"
- *Arabis mosaic virus small satellite RNA*
- *Cereal yellow dwarf virus-RPV satellite RNA*
- *Chicory yellow mottle virus satellite RNA*
- *Lucerne transient streak virus satellite RNA*
- *Solanum nodiflorum mottle virus satellite RNA*
- *Subterranean clover mottle virus satellite RNA*
- *Tobacco ringspot virus satellite RNA*
- *Velvet tobacco mottle virus satellite RNA.*

Vault RNA

ault RNAs (vRNA) are an RNA family found as part of the *vault ribonucleoprotein complex* first discovered in 1986. The complex consists of a *major vault protein* (MVP), two *minor vault proteins* (VPARP and TEP1), and several small untranslated RNA molecules. Each vault particle contains 8-16 vRNA molecules. The vault complex has been linked to drug resistance. Cryo-electron microscopy has revealed the vRNA resides internally close to the ends of the vault caps. The position of vRNA suggests that it could interact with both the interior and exterior of the vault particle. It has been shown that TEP1 is involved in stabilising the vRNA.

Species Distribution

Vault RNAs have been found in:

- Human
- Rodents
- Bullfrog.

Vaults but not the vRNA have also been isolated from:

- sea urchin
- Dictyostelium discoideum
- Acanthamoeba.

Role in Multidrug Resistance

Vault RNAs, in conjunction with the vault ribonucleoprotein complex, have long been connected with drug resistance, though the mechanisms involved have only recently been determined. Vault non-coding RNAs produce small vault RNAs (svRNAs) through a DICER mechanism which then operate in similar manner to miRNAs: an svRNA binds an argonaute protein and downregulates expression of CYP3A4, an enzyme involved in drug metabolism.

In Translation

Messenger RNA (mRNA) carries information about a protein sequence to the ribosomes, the protein synthesis factories in the cell. It is coded so that every three nucleotides (a codon) correspond to one amino acid. In eukaryotic cells, once precursor mRNA (pre-mRNA) has been transcribed from DNA, it is processed to mature mRNA. This removes its introns—non-coding sections of the pre-mRNA. The mRNA is then exported from the nucleus to the cytoplasm, where it is bound to ribosomes and translated into its corresponding protein form with the help of tRNA.

In prokaryotic cells, which do not have nucleus and cytoplasm compartments, mRNA can bind to ribosomes while it is being transcribed from DNA. After a certain amount of time the message degrades into its component nucleotides with the assistance of ribonucleases. Transfer RNA (tRNA) is a small RNA chain of about 80 nucleotides that transfers a specific amino acid to a growing polypeptide chain at the ribosomal site of protein synthesis during translation. It has sites for amino acid attachment and an anticodon region for codon recognition that binds to a specific sequence on the messenger RNA chain through hydrogen bonding.

Ribosomal RNA (rRNA) is the catalytic component of the ribosomes. Eukaryotic ribosomes contain four different rRNA molecules: 18S, 5.8S, 28S and 5S rRNA. Three of the rRNA molecules are synthesized in the nucleolus, and one is synthesized elsewhere. In the cytoplasm, ribosomal RNA and protein combine to form a nucleoprotein called a ribosome. The ribosome binds mRNA and carries out protein synthesis. Several ribosomes may be attached to a single mRNA at any time. rRNA is extremely abundant and makes up 80% of the 10 mg/ml RNA found in a typical eukaryotic cytoplasm.

Transfer-messenger RNA (tmRNA) is found in many bacteria and plastids. It tags proteins encoded by mRNAs that lack stop codons for degradation and prevents the ribosome from stalling.

Regulatory RNAs

Several types of RNA can downregulate gene expression by being complementary to a part of an mRNA or a gene's DNA. MicroRNAs(miRNA; 21-22 nt) are found in eukaryotes and act through RNA interference (RNAi), where an effector complex of miRNA and enzymes can break down mRNA which the miRNA is complementary to, block the mRNA from being translated, or accelerate its degradation. Whilesmall interfering RNAs (siRNA; 20-25 nt) are often produced by breakdown of viral RNA, there are also endogenous sources of siRNAs. siRNAs act through RNA interference in a fashion similar to miRNAs. Some miRNAs and siRNAs can cause genes they target to bemethylated, thereby decreasing or increasing transcription of those genes. Animals have Piwi-interacting RNAs (piRNA; 29-30 nt) which are active in germline cells and are thought to be a defence against transposons and play a role in gametogenesis.

Many prokaryotes have CRISPR RNAs, a regulatory system similar to RNA interference. Antisense RNAs are widespread; most downregulate a gene, but a few are activators of transcription. One way antisense RNA can act is by binding to an mRNA, forming double-stranded RNA that is enzymatically degraded. There are many long noncoding RNAs that regulate genes in eukaryotes, one such RNA is Xist which coats one X chromosome in female mammals and inactivates it. An mRNA may contain regulatory elements itself, such as riboswitches, in the 5' untranslated region or 3' untranslated region; these cis-regulatory elements regulate the activity of that mRNA. The untranslated regions can also contain elements that regulate other genes.

In RNA Processing

Many RNAs are involved in modifying other RNAs. Introns are spliced out of pre-mRNA byspliceosomes, which contain several small nuclear RNAs (snRNA), or the introns can be ribozymes that are spliced by themselves. RNA can also be altered by having its nucleotides modified to other nucleotides than A, C, G and U. In eukaryotes, modifications of RNA nucleotides are generally directed by small nucleolar RNAs (snoRNA; 60-300 nt), found in the nucleolusand cajal bodies. snoRNAs associate with enzymes and guide them to a spot on an RNA by basepairing to that RNA. These enzymes then perform the nucleotide modification. rRNAs and tRNAs are extensively modified, but snRNAs and mRNAs can also be the target of base modification.

RNA Genomes

Like DNA, RNA can carry genetic information. RNA viruses have genomes composed of RNA, and a variety of proteins encoded by that genome. The viral genome is replicated by some of those proteins, while other proteins protect the genome as the virus particle moves to a new host cell.Viroids are another group of pathogens, but they consist only of RNA, do not encode any protein and are replicated by a host plant cell's polymerase.

In Reverse Transcription

Reverse transcribing viruses replicate their genomes by reverse transcribing DNA copies from their RNA; these DNA copies are then transcribed to new RNA. Retrotransposons also spread by copying DNA and RNA from one another, and telomerase contains an RNA that is used as template for building the ends of eukaryotic chromosomes.

Double-stranded RNA

Double-stranded RNA (dsRNA) is RNA with two complementary strands, similar to the DNA found in all cells. dsRNA forms the genetic material of some viruses (double-stranded RNA viruses). Double-stranded RNA such as viral RNA or siRNA can trigger RNA interference ineukaryotes, as well as interferon response in vertebrates.

Key Discoveries in RNA Biology

Research on RNA has led to many important biological discoveries and numerous Nobel Prizes. Nucleic acids were discovered in 1868 by Friedrich Miescher, who called the material 'nuclein' since it was found in the nucleus. It was later discovered that prokaryotic cells, which do not have a nucleus, also contain nucleic acids. The role of RNA in protein synthesis was suspected already in 1939. Severo Ochoawon the 1959 Nobel Prize in Medicine (shared with Arthur Kornberg) after he discovered an enzyme that can synthesize RNA in the laboratory. Ironically, the enzyme discovered by Ochoa (polynucleotide phosphorylase) was later shown to be responsible for RNA degradation, not RNA synthesis.

The sequence of the 77 nucleotides of a yeast tRNA was found by Robert W. Holley in 1965, winning Holley the 1968 Nobel Prize in Medicine (shared with Har Gobind Khorana and Marshall Nirenberg). In 1967, Carl Woese hypothesized that RNA might be catalytic and suggested that the earliest forms of life (self-replicating molecules) could have relied on RNA both to carry genetic information and to catalyse biochemical reactions—an RNA world.

During the early 1970's, retroviruses and reverse transcriptase were discovered, showing for the first time that enzymes could copy RNA into DNA (the opposite of the usual route for transmission of genetic information). For this work, David Baltimore, Renato Dulbecco and Howard Temin were awarded a Nobel Prize in 1975. In 1976, Walter Fiers and his team determined the first complete nucleotide sequence of an RNA virus genome, that of bacteriophage MS2.

In 1977, introns and RNA splicing were discovered in both mammalian viruses and in cellular genes, resulting in a 1993 Nobel to Philip Sharpand Richard Roberts. Catalytic RNA molecules (ribozymes) were discovered in the early 1980s, leading to a 1989 Nobel award to Thomas Cech and Sidney Altman. In 1990 it was found in petunia that introduced genes can silence similar genes of the plant's own, now known to be a result of RNA interference. At about the same time, 22 nt long RNAs, now called microRNAs, were found to have a role in the development of *C. elegans*. Studies on RNA interference gleaned a Nobel Prize for Andrew Fire and Craig Mello in 2006, and another Nobel was awarded for studies on transcription of RNA to Roger Kornberg in the same year. The discovery of gene regulatory RNAs has led to attempts to develop drugs made of RNA, such as siRNA, to silence genes.

Protein Biosynthesis

Protein synthesis is the process in which cells build proteins. The term is sometimes used to refer only to protein translation but more often it refers to a multi-step process, beginning with amino acid synthesis and transcription of nuclear DNAinto messenger RNA, which is then used as input to translation.

The cistron DNA is transcribed into a variety of RNA intermediates. The last version is used as a template in synthesis of a polypeptide chain. Proteins can often be synthesized directly from genes by translating mRNA. When a protein needs to be available on short notice or in large quantities, a protein precursor is produced. Aproprotein is an inactive protein containing one or more inhibitory peptides that can be activated when the inhibitory sequence is removed by proteolysis duringposttranslational modification. A preprotein is a form that contains a signal sequence (an N-terminal signal peptide) that specifies its insertion into or through membranes; i.e., targets them for secretion. The signal peptide is cleaved off in theendoplasmic reticulum. Preproproteins have both sequences (inhibitory and signal) still present.

For synthesis of protein, a succession of tRNA molecules charged with appropriate amino acids have to be brought together with an mRNA molecule and matched up by base-pairing through their anti-codons with each of its successive codons. The amino acids then have to be linked together to extend the growing protein chain, and the tRNAs, relieved of their burdens, have to be released. This whole complex of processes is carried out by a giant multimolecular machine, the ribosome, formed of two main chains of RNA, called ribosomal RNA (rRNA), and more than 50 different proteins. This molecular juggernaut latches onto the end of an mRNA molecule and then trundles along it, capturing loaded tRNA molecules and stitching together the amino acids they carry to form a new protein chain.

Protein biosynthesis, although very similar, is different for prokaryotes and eukaryotes.

6

DNA Quantification and Quality Assessment

DNA quantification and quality assessment is done by using UV-VISspectrophotometry.Normally quality check is performed through the A260/ A280 ratio that is 1.8 value shows the highest purity, if more than 1.8 shows the presence of RNA contamination and less than that shows the protein contamination (Sambrook, *et. al,* 1989).

AGE is conducted of isolated DNA samples for visualization of bands. For perfect extraction, there will be clear single bands otherwise smear like image will be seen. That is in case of degraded DNA. For PCR based methods the DNA samples are diluted for PCR requirements. The optimum concentration is usually 40- 60ng/mL.

Polymerase Chain Reaction (PCR)

The DNA amplification by thermal cycling called Polymerase Chain Reaction is in vitro method that can be used to amplify a specific DNA segment from small amounts of DNA template or duplex into millions of copies. It is invented by Kary Mullis *et.al,* (1985).

Heatdenaturation

This temperature denatures the double stranded DNA into two individual stands. Denaturation temperature is 95^0 C for 30 seconds or is 97^0 C for 15 seconds, however higher temperature may be appropriate, especially for G+C rich nucleotides.

Annealing

During this time one primer binds with the 5 prime end of one DNA strand and the other primer binds with 3 prime end of its complementary strand. Annealing is hybridization of primers to single stranded DNA and the length of time required for primer annealing depends on the basic composition, length and concentration of primers.

Primer Extension

This temperature is varies for Taq DNA polymerase which adds complementary nucleotides one by one to the 3' OH group of the primer. Estimates for the rate of nucleotide incorporation at 72^0 C vary from 35-100 nucleotides per second depending upon the buffer, pH salt concentration and nature of DNA template.

Number of PCR Cycles

The optimum number of cycles depends mainly upon the starting concentration of target DNA, when other parameters are optimized. A common mistake is to execute too many cycles. The increase in the number of cycles will increase the amount and complexity of non-specific background products.

Restriction Fragment Length Polymorphisms (RFLPs)

RFLP Definition

The variation(s) in the length of DNA fragments produced by a specific restriction endonuclease from genomic DNAs of two or more individuals of a species (Kahl, 2001).

Restriction fragment length polymorphisms (RFLPs) was the first to develop (Soller and Beckmann, 1983) and are still widely used marker system especially in practical plant breeding. It was first developed in the 1980s for use in human genetic applications and was later applied to plants Most of the studies in systematic and evolution (Whitkus *et al.*, 1994) and molecular maps of different crop species were initially made through the use of RFLP markers. Restriction fragment length polymorphism (RFLP) technology. RFLPs are relatively small in size and are co-dominant in nature.

Restriction fragments of different lengths are thus generated. All RFLP markers are analysed using a common technique. However, the analysis requires a relatively complex technique that is time consuming and expensive.

The hybridization results can be visualized by autoradiography (if the probes are radioactively labelled), or using chemiluminesence (if non-radioactive, enzyme-linked methods are used for probe labelling and detection). Any of the visualization techniques will give the same results. The visualization techniques used will depend on the laboratory conditions. A comprehensive study on these maps is documented by Phillips and Vasil (1994). Thus, RFLP markers have the potential to address with power, precision and accuracy, all issues of plant breeding in which markers can be applied.

In addition to that, voluminous literature is available on utilization of RFLPs in other aspects especially in DNA fingerprinting (Tanksley, 1983), phylogenetic analysis (Zheng *et al.*, 1994), detection of genetic variability in rice, genetic dissection of maturity (Phillips *et al.,* 1992), identification of chromosomal regions responsible for regeneration of RFLPs, comparative mapping of chromosomes in wheat, rye and barley, analysis of wide compatibility in rice varieties (Zheng *et al.*, 1994), QTL mapping, gene tagging, and identification of ecotypes (Pupilli *et al.,* 2000).

To discover RFLPs, restriction enzymes (RE) are used to cut DNA at specific 4-6 bp recognition sites. Sample DNA is digested with one or more RE's and resulting fragments are separated according to molecular size using gel electrophoresis. After that Southern Blotting is performed where fragments are immobilized on a nylon membrane, followed by hybridization to a labelled probe (specific chemoluminescence DNA probe) that identifies the locus under autoradiograph.

Fragment sizes of digested DNA will differ depending on the presence or absence of the proper recognition sequence for the enzyme. Differences in fragment length result from base substitutions, additions, deletions or sequence rearrangements within RE recognition sequences. Although two individuals of the same species have almost identical genomes, they will always differ at a few nucleotides.

Some of these differences will produce new restriction sites (or remove them), and the banding pattern seen on a genomic Southern will thus be affected. For any given probe (or gene), it is often possible to test different restriction enzymes until one which gives a pattern difference between two individuals is found, that is, a RFLP. The less related the individuals, the more divergent their DNA sequences are and the more likely you are to find a RFLP.

However, the low level of polymorphism present in RFLPs restricted the use of this technique. In addition, RFLP requires large quantities of relatively pure DNA, and the frequent use of radioisotopes in detection step makes it technically demanding, laborious and costly to characterize large number of samples.

Utilization of RFLP markers requires a lot of skill in using short-lived radioactive material: the availability of which often becomes a limiting factor especially in the developing countries. Another limiting factor could be the large sample size to be analysed like that of F segregating population. In this case, utilization of RFLP markers may not be feasible, as it requires extraction of DNA in ultra pure form, restriction digestion, and lengthy and expensive process of isotopic labeling. Such difficulties have opened up a debate on the feasibility of routine application of RFLPs in large scale crop improvement programme. Another difficulty frequently encountered by RFLP users is the detection of low (1-3 loci) ornon-existent polymorphism in closely related species/varieties (Waugh and Powell, 1992). It was essentially this difficulty that led to the search for a new marker system.

Randomly Amplified Polymorphic DNAs (RAPDs)

Randomly-amplified polymorphic DNA markers (RAPD). In 1991 Welsh and McClelland developed a new PCR-based genetic assay namely randomly amplified polymorphic DNA (RAPD). This procedure detects nucleotide sequence polymorphisms in DNA by using a single primer of arbitrary nucleotide sequence. In this reaction, a single species of primer anneals to the genomic DNA at two different sites on complementary strands of DNA template. If these priming sites are within an amplifiable range of each other, a discrete DNA product is formed through thermocyclic amplification. On an average, each primer directs amplification of several discrete loci in the genome, making the assay useful for efficient screening of nucleotide sequence polymorphism between individuals.

However, due to the stoichastic nature of DNA amplification with random sequence primers, it is important to optimize and maintain consistent reaction conditions for reproducible DNA amplification. RAPDs are DNA fragments amplified by the Polymerase Chain Reaction (PCR) using short (generally 10 bp) synthetic primers of random sequence. Because of the short primers used, the reannealing temperature in the PCR must be low (35-40°C) for the primer to bind. However, due to low temperature, the binding is not very specific, which means that primers will bind also to sequences which are not completely complementary.

These oligonucleotides serve as both forward and reverse primer and usually are able to amplify fragments from 3-10 genomic sites simultaneously. Amplified fragments (within the 0.5-5 kb range) are separated by gel-electrophoresis and polymorphisms are detected as the presence or absence of bands of particular size. These polymorphisms are considered to be primarily due to variation in the primer annealing sites.

They are dominant markers and hence have limitations in their use as markers for mapping, which can be overcome to some extent by selecting those markers that are linked in coupling. RAPD assay has been used by several groups as efficient tools for identification of markers linked to agronomically important traits, which are introgressed during the development of near isogenic lines. The application of RAPDs and their related modified markers in variability analysis and individual-specific genotyping has largely been carried out, but is less popular due to problems such as poor reproducibility faint or fuzzy products, and difficulty in scoring bands, which lead to inappropriate inferences.

In RAPD analysis the total DNA from an organism is mixed with (generally) 10 bp single stranded DNA (primer) together with the four different deoxynucleotides and a heat stable DNA polymerase enzyme. The reaction mixture is placed on a thermocycler (PCR-machine) which can change the temperature of the reaction rapidly according to a predefined PCR programme.

Reannealing at low temperature makes the primer molecules attach to complementary site on the DNA. Elongation is the period when the polymerase enzyme elongates the attached primers from their 3' end. Denaturation separates the new formed strands.

After a few PCR cycles DNA pieces with well defined length similar to the one between the two original primer sites dominate the mixture. This is because they amplify exponentially. In most RAPD reactions the result will be amplification of 5 to 10 different DNA segments whose lengths depend on the primer recognition sites available in the genome. The DNA after amplification is visualized for bands using Agarose Gel Electrophoresis (AGE). The band scoring is done by comparing with the marker bands.

Inter Simple Sequence Repeats (ISSR)

ISSR amplification definition: A variant of the polymerase chain reaction that uses simple sequence repeat primers (e.g. [AC] n) to amplify regions between their target sequences (Kahl, 2001).

Inter simple sequence repeat markers (ISSR). In this technique, reported by Zietkiewicz et al. primers based on microsatellites are utilized to amplify inter-SSR DNA sequences. Here, a single primer or combination of primers composed of a microsatellites sequence anchored with a single base at 3' region so that it can amplify regions between two microsatellites loci with oligonucleotides annealing to repeat sequence.

More complex banding patterns can be achieved using 5'-anchored primers that incorporate the SSR regions in their amplification products, and by combining 3'- and 5'- primers.ISSR circumvents the requirement for flanking sequence information, and thus has found wide applicability in a variety of plants.(Wolfe et al.1998} These are mostly dominant markers, though occasionally a few of them exhibit codominance.

An unlimited number of primers can be synthesized for various combinations of di-, tri-, tetra- and pentanucleotides [(4)3 = 64, (4)4 = 256] etc. with an anchor made up of a few bases and can be exploited for a broad range of applications in plant species. ISSR analysis is technically simpler than any other marker systems.

The method provides highly reproducible results and generates higher levels of polymorphism compared to other markers in many systems. The advantage lies in long primer length and stringent annealing temperature. Primers used (from Yang *et al.,* 1996): For separation and visualization of bands Agarose gel electrophoresis is conducted.

Simple Sequence Repeats (SSRs)

SSR (Microsatellites) definition: Any one of a series of very short (2-10 bp), middle repetitive, tandemly arranged, highly variable (hyper variable) DNA sequences dispersed throughout fungal, plant, animal and human genomes (Kahl, 2001).

Simple sequence repeats (SSR) or microsatellites are a class of repetitive DNA elements. The di-, tri- or tetra-nucleotide repeats are arranged in tandem arrays consisting of 5 – 50 copies, such as (AT) 29, (CAC) 16 or (GACA) 32. SSRs are abundant in plants, occurring on average every 6-7 kb (Cardle *et al.*, 2000).

These repeat motifs are flanked by conserved nucleotide sequences from which forward and reverse primers can be designed to PCR-amplify the DNA section containing the SSR.SSR alleles, amplified products of variable length, can be separated by gel electrophoresis and visualized by silver-staining, autoradiography (if primers are radioactively labelled) or via automation (if primers are fluorescently labelled).

SSR analysis is amenable to automation and multiplexing, and allows genotyping to be performed on large numbers of lines, and multiple loci to be analysed simultaneously. SSRs can be identified by searching among DNA databases (e.g. EMBL and Gene bank), or alternatively small insert (200-600bp) genomic DNA libraries can be produced and enriched for particular repeats (Powell *et al.*, 1996). From the sequence data, primer pairs (of about 20 bp each) can be designed (software programmes are available for this).

For separation and visualization of bands Polyacrylamide Gel Electrophoresis (PAGE) is conducted (6% acrylamide solution), stained and the observed the bands.

Amplified Length Polymorphisms (AFLPs)

AFLP definition: Any difference between corresponding DNA fragments from two organisms A and B that is detected by the amplified restriction length polymorphism technique (Kahl, 2001).

Table: *DNA Fingerprinting Techniques: A Comparative Analysis*

Marker	*Advantages*	*Disadvantages*
RFLP	Unlimited number of loci	Labour intensive
	Co-dominant	Fairly expensive
	Many detection systems	Large quantity of DNA needed
	Can be converted to SCARs	Often very low levels of polymorphism
	Robust in usage	Can be slow (often longexposure times)
	Good use of probes fromother species	Needs considerable degree ofskill
	Detects in related genomes	No sequence informa-tionrequired
RAPD	Results obtained quickly	Highly sensitive to laboratorychanges
	Fairly cheap	Low reproducibility within andbetween laboratories
	No sequence informationrequired	Cannot be used acrosspopulations nor across species
	Relatively small DNAquantities required	Often see multiple loci
	High genomic abundance	Dominant
	Good polymorphism	Can be automated
ISSR	Highly polymorphic	Usually dominant
	Robust in usage	Species-specific
SSR	Can be automated	Fast High developmental andstartup costs
	Highly polymorphic	Species-specific
	Robust	Sometimes difficult interpretation because of stuttering
	Can be automated	Usually single loci even in polyploids
	Only very small DNA	<!—[if!support Misaligned Columns]—> <!—[endif]—>

Marker	*Advantages*	*Disadvantages*
AFLP	Small DNA quantitiesrequired	Evaluation of up to 100 loci
	No sequence informationrequired	Marker clustering
	Can be automated	Dominant
	Can be adapted for differentuses, e.g.	Technique is patented
cDNA-AFLP	Can be technically challenging	

This genetic marker system basically uses restriction fragment analysis, but a PCR approach is used to visualize fragments instead of hybridization with a probe. Because of the PCR step, this type of markers has characteristics quite different from RFLP markers. Initially, DNA is extracted as for RFLP and then cut with two different restriction enzymes to produce well defined restriction fragments with sticky ends. Synthetic double stranded linkers of approximately 18 to 20 bp with matching sticky ends are ligated on all the restriction fragments. Ligated fragments are subsequently amplified in PCR with 18 to 20 nucleotide length primers recognizing linkers in each end of the fragments. A primer used for amplification in AFLP is normally labelled with P^{33} or some nonradioactive labeling system.

The mixture of amplified fragments are separated according to size in a Polyacrylamide gel and visualized by means of autoradiography or other procedures for development of non radioactive systems. Alternatively, automated sequencing machines can be used to read the sequence of amplification products directly from the Polyacrylamide gel if primers have been labelled with special fluorescent dyes. Also, silver staining of the Polyacrylamide gel after electrophoresis can be used to stain all amplified DNA fragments in the gel. To avoid that all restriction fragments from the genome amplify in the same PCR, which will produce a smear because of too many DNA fragments, a two step amplification procedure is used. Firstly, part of the total number of restriction fragments are amplified in the "preamplification" with primers containing one extra "selective" nucleotide on their 3' end.

This selective nucleotide will allow amplification only of restriction fragments with a matching nucleotide next to the linker. With one selective nucleotide on both primers only 1/16th of all restriction fragments in the mixture will amplify during the preamplification. During the subsequent "selective" amplification, additionally, one or two selective nucleotides on each primer will further reduce the number of fragments amplified. A good AFLP amplification will show 90 to 100 different fragments in one analysis.

Applications

There are a variety of applications in DNA fingerprinting, such as to identify genetic diversity within breeding populations in plants and animals, to differentiate between plant species cultivars, as well as to identify plants containing a gene of interest. , it is also used to detect genetically modified organisms in agriculture in the United States.

RFLP is most suited to studies at the intraspecific level or among closely related taxa. Presence and absence of fragments resulting from changes in recognition sites are used for identifying species or populations. . In plant genetic analysis, RFLP markers were initially used for estimating genetic distance and fingerprinting in wheat. In plant and animal breeding, they can reveal which backcross progeny most resemble the desired parent. RFLPs are used to settle cases where biological parentage is at issue.

Many plant species are being analysed using RAPD method. RAPD markers were also used for characterisation, estimation of genetic relatedness and determination of genetic diversity of tea germplasm. These markers are able to distinguish between genotypes but limited to comparisons of populations from a few sources. Lastly RAPD analysis was used to assess genetic differences and relatedness among fig, *Ficus carica* L. clones.

For AFLP analysis, it is an efficient technology for assessing genetic diversity among cultivars such as wheat, *Triticum aestivum* cultivars.. Cultivars were improved through crossing with elite breed however, crossing within such germplasm may lead to a reduction in the level of genetic variation among parental lines due to intense selection pressure.

Thus DNA marker systems are used to reflect actual DNA differences among lines since selection pressure and genetic drift are accounted for. It is an efficient genetic marker system as the technique gives a much higher degree of polymorphism than RAPD or RFLP. In addition, its prior sequence knowledge is not required as the fragments detected are selected by specific primer sets in the PCR reaction. It is reliable as stringent reaction conditions are applied for primer annealing.

Inter SSR (ISSR) fingerprinting was developed such that no sequence knowledge was required. Primers based on a repeat sequence, such as (CA) n, can be made with a degenerate 3'-anchor, such as (CA) 8RG or (AGC) 6TY. The resultant PCR reaction amplifies the sequence between two SSRs, yielding a multilocus marker system useful for fingerprinting, diversity analysis and genome mapping. Previous investigators have demonstrated that ISSR analysis usually detects a higher level of polymorphism than that detected with restriction fragment length polymorphism (RFLP) or random amplified polymorphic DNA (RAPD) analyses.

As with AFLPs, the great advantage of SSRs analysis is the large number of polymorphisms that the method reveals, with one locus in soybean (Glycine max) reported to have 26 alleles. Unlike AFLPs, microsatellites are co-dominant markers, thus heterozygotes can be readily identified. Microsatellite co-dominance will increase the efficiency and accuracy of population genetic measures based on these markers compared with other markers, such as AFLPs and RAPDs.

With a beginning as a mere forensic tool, the world of DNA fingerprinting has a gone a long way in revealing the genetic identity of living beings. With massive evolution in concepts and techniques it has given a lot to various fields cutting across the spectrum. In plants it has not only helped in identifying species but also in defining a new realm in plant genomics, plant breeding and in conserving the biodiversity. With world paving way for developments in biotechnology, DNA fingerprinting promises a very powerful tool in our future endeavors.

Intellectual Property Rights in Plant Varieties

Legal and Policy Concepts Relating to the Protection of Plant Varieties and Plant Breeders' Rights

Intellectual property rights (IPRs) are legal rights granted by governmental authorities to control certain products of human intellectual effort and ingenuity. (OECD 1996). An in-depth discussion of the philosophical and policy goals served by granting legal protection to these products is beyond the scope of this study.

However, a basic familiarity with these goals is necessary to grasp how national and international intellectual property systems and institutions have evolved into their present forms and to understand the constraints that those systems and institutions place on governments seeking to implement competing policy objectives in tension with IPRs.

Two broad philosophical approaches underlie the decision to grant IPRs in the products of human intellectual effort and ingenuity. Elements of these two approaches can be found to different degrees in all national laws and international agreements relating to IPRs.

The first approach to IPR protection predominates in many civil law legal systems, including continental Europe. It takes the position that the products of the human mind are stamped with the personality of their creator, inventor or author, thus endowing him or her with a moral as well as an economic claim to exploit those products to the exclusion of third parties. Under this view, legal protection flows from a state's commitment to protect human rights, a fact reflected in the wording of article 27 of the Universal Declaration of Human Rights, which guarantees to everyone "the right to the protection of the moral and material interests resulting from any scientific, literary or artistic production of which he is the author." Legal protection also rewards creators, inventors and authors for their intellectual efforts as well as their expenditures of time and money. From these underlying premises flows a desire to provide robust intellectual property protection that includes, for example, a broad and expansive range of exclusive rights, long terms of protection, limits on mandatory licensing and narrow exceptions and limitations to exclusive rights.

The second approach to IPR protection takes as its starting premise an instrumental view of intellectual property. Legal protection for the products of human intellectual effort and ingenuity is granted not because of a moral commitment to compensating creators or innovators, but rather because the products they create enrich a society's culture and knowledge and thus increase its welfare. Perhaps the most well-known manifestation of this approach is found in the Intellectual Property Clause of the United States Constitution, which authorizes the United States Congress "[t]o promote the progress of science and useful arts, by securing for limited times to authors and inventors the exclusive right to their respective writings and discoveries."

This instrumentalist philosophy shapes the structure of many national intellectual property systems. The grant of IPRs in nations following an instrumentalist approach is intended to provide adequate incentives for creators, inventors and authors to invest the time, resources and intellectual capital needed to create intellectual property products.

In the absence of a grant of exclusive rights over those products, so-called "free riders" who have not made such investments could exploit inexpensive distribution and reproduction technologies and sell others' intellectual property products at a much lower cost. However, the ultimate goal of legal protection is not remunerative reward for creators but the enhancement of social welfare through access to the ideas and information contained in their products. Instrumentalist intellectual property systems often tailor the scope of legal protection to achieve this goal, for example by placing certain limits on the scope of protection or recognizing situations in which consumers or second-generation creators may access and use intellectual property products for socially valuable purposes.

Policy Objectives Favouring IPRs in New Plant Varieties

The policy goals of granting IPRs to plant varieties are grounded principally on an instrumentalist approach to IPRs. This is true both for patents and plant breeders' rights.

Under this instrumentalist approach, new plant varieties are afforded legal protection to encourage commercial plant breeders to invest the resources, labour and time needed to improve existing plant varieties by ensuring that breeders receive adequate remuneration when they market the propagating material of those improved varieties. In the absence of a grant of exclusive rights to breeders, the dangers of free riding by third parties would be considerable.

This is because the genetic material within plants that specifies their distinctive and commercially valuable features is naturally self-replicating, for example by reproduction of seeds or other propagating material. Self-replication makes innovations incorporating biological material particularly susceptible to exploitation by parties other than the innovator.

IPRs in plant varieties thus provide some assurance to breeders that they will be able to recoup the risks and costs of a value-added innovation that is based upon an underlying biological resource. Ultimately, however, the grant of exclusive rights to plant breeders is designed to benefit the society granting the rights. It provides an incentive for private research and development into new breeding techniques, thereby reducing the need for government funding to subsidize these activities. It encourages the development of new and beneficial plant varieties for use by farmers and consumers. And it furthers the society's development of agriculture, horticulture and forestry.

An international system of IPR protection for plant varieties expands these benefits by facilitating access to new varieties created in other states. Once breeders are assured that their rights will be protected in other states, breeders will be more willing to make their new varieties available in those states (assuming they have access to a distribution and marketing infrastructure). This benefits farmers, consumers and researchers in many more jurisdictions.

The Evolution and Structure of the International Intellectual Property System

The different policy objectives underlying the protection of IPRs have shaped the structure and evolution of the international intellectual property system. Most early domestic intellectual property laws provided no legal protection to intellectual property products created in other nations, thereby permitting those products to be exploited by free riders operating outside the state in which the products were created. The unfairness of this result prompted governments in the late nineteenth century to consider an international approach to protect IPRs.

Limited treaty obligations: The drafters of the first multilateral intellectual property treaties quickly realized, however, that there was insufficient political support for reconciling many of the differences that existed among national IPR laws. For this reason, the drafters abandoned the idea of harmonizing diverse national laws to create a single, international IPR applicable in all signatory states. They fashioned instead a system that creates a limited set of treaty-based obligations that each member state of that system is then required to implement in its national IPR laws.

Implementation of treaty-based obligations in national IPR laws can occur in one of two ways. In some nations (often referred to as "automatic incorporation" states), treaties become binding as a part of domestic law as soon as formal ratification procedures have been adopted.

In these nations, treaties are considered to be "self-executing" or capable of being given "direct effect" in domestic law such that courts and administrative agencies can construe the treaty directly and enforce the rights it grants to the owners of intellectual property products. In other nations, however (often referred to as "legislative incorporation" states), treaties are considered to be "non-self-executing" and can only become binding in domestic law once the parliament or legislature has adopted legislation to implement the treaty. In these nations, owners of intellectual property products rely on this domestic legislation rather than on the treaties themselves when they seek to enforce rights granted to them under the treaties.

The territoriality of IPRs: Because of the limited scope of international IPR agreements, there are to this day (with the limited exception of the European Union) no *international* IPRs available to inventors and creators who seek to market their products across borders; rather, IPRs are territorial in nature and are acquired and enforced on a country-by-country basis under territorially-circumscribed *national* IPR laws.

Thus, for example, the inventor of a genetically enhanced variety of corn who seeks patent protection for that variety must apply for protection in each country in which he or she hopes to sell the corn. The inventor must comply with all of the requirements that each country imposes for granting patent rights to the new variety.

Similarly, once protection is granted, issues such as the scope of the exclusive rights the inventor enjoys in the patented plant variety, the term of patent protection and the limitations imposed on the inventor's rights are all determined by the different national laws. Recent international agreements have achieved some modest forms of procedural harmonization, but they have not altered the fundamental premise that national laws rather than international treaties are the immediate source of nearly all private rights in intellectual property products.

Two basic principles flow from this territorial approach to protecting IPRs. First, where national laws differ as to the scope or content of the protection they provide to intellectual property products, the rights enjoyed by the owners of those products will vary in different national jurisdictions. Second, territoriality implies that each nation has the right to decide on the form of IPR protection to be granted within its own borders, provided that it complies with the obligations contained in international IPR agreements to which it is a party.

Although territoriality thus gives governments some autonomy to set national IPR policies within their own borders, states often view the policies other governments choose as a subject of concern. Indeed, the global reach of markets for intellectual property products makes this concern a necessity.

To take just one example, most patent laws grant inventors owning patents within a state the right to prevent the importation into that state of products created in other nations that contain the patented invention. Thus, where distribution markets transcend national borders, an industry may find itself precluded from distributing products in other jurisdictions as a result of patent rights, as occurred when Indian cotton producers were precluded from importing certain forms of transgenic cotton into the United States.

Core obligations imposed by international intellectual property agreements: The territorial approach to IPR protection appears at first to present myriad difficulties for creators and owners of intellectual property products. In fact, however, the content of each nation's IPR laws are often quite similar since they have been shaped by international IPR agreements ratified by many states. In addition, the obligations these agreements impose have expanded over time, thus narrowing (although by no means eliminating) the differences among national intellectual property systems. The following sections briefly explain the core obligations contained in most international IPR agreements. More detailed information concerning the rights and obligations contained in specific IPR agreements relating to plant varieties is provided in Part II below.

National treatment: One of the cornerstones of international IPR agreements is the national treatment principle. National treatment bars discrimination against foreign IPR owners by requiring that each state provide the same IPRs to private parties from other member states as are provided to the state's own nationals. National treatment levels the playing field among treaty parties and prevents a state from giving its own creators and inventors unfair advantages over foreign creators and inventors. In the absence of national treatment, for example, domestic firms could freely exploit intellectual property products created in other member states while simultaneously enjoying legal protection within their own domestic markets.

***Reciprocity*:** The provisions of several intellectual property treaties contain a limited exception to national treatment known as reciprocity. Where a treaty permits reciprocity, State A may condition the grant of legal protection to intellectual property products from State B upon State B's granting of legal protection to intellectual property products from State A. Reciprocity is often applied to new IPRs as means of encouraging other nations to recognize the new rights and extend their protection to foreign nationals. Once a large number of states have recognized the new IPR, they may revise the treaty to eliminate the reciprocity option and impose a national treatment obligation.

***Most favoured nation treatment*:** The most favoured nation ("MFN") principle is a common feature of international trade agreements but has only recently been applied to IPRs. The principle extends the national treatment rule by compelling a government that provides a privilege or benefit to *one* state within a treaty system automatically to grant that same privilege or benefit to *all* states within the same system. The MFN principle thus prevents a subset of states within a larger treaty system from entering into bilateral or other special agreements among themselves, unless they grant the rights contained in those agreements to all other parties within the larger treaty system.

***Subject matter and eligibility requirements*:** Intellectual property agreements specify the subject matter characteristics of the products that are eligible for legal protection. In the context of patents, for example, a treaty may specify the types of inventions (such as products and processes) to which states parties must grant legal rights. These subject matter requirements are generally drafted using language that instructs member states concerning the basic characteristics that a product must possess for it to merit protection under domestic IPR laws, while preserving sufficient flexibility for states to tailor the details of protection to the particularities of their national legal systems.

***Exclusive rights*:** Where an intellectual property product satisfies a treaty's subject matter eligibility requirements, states parties are required to grant an enumerated set of exclusive rights with respect to that product. These exclusive rights grant to the product's owner the power to exclude all third parties from engaging in the activity that the right covers (such as reproducing or modifying the product or distributing it to others). It is the exclusivity of the rights granted that allows IPR owners to recoup the investment of time, money and resources required to create intellectual property products. The particular exclusive rights mandated by IPR agreements differ depending on the specific type of IPR product at issue. They are drafted with greater or lesser degrees of specificity depending on the degree of consensus among member states' national IPR laws, thus affording varying levels of discretion to governments to implement the rights in their domestic legal systems.

***Terms of protection and the public domain*:** Intellectual property agreements also specify for their states parties the minimum term of years during which intellectual property products must receive legal protection. Once that term has expired, the treaties do not require states to grant legal protection to the products. Thus, unless the state adopts a longer term of protection, after the expiration of the initial term of protection the product may be freely used by anyone for any purpose, including as a source for creating new products or simply for consumption. A corollary of this rule is that national IPR laws do not permit putative inventors and creators to claim IPRs in materials as they are found in nature or where they are already part of the public domain.

***Exceptions and limitations to exclusive rights*:** International IPR agreements constrain the ability of national governments seeking to restrict the exercise of IPRs to achieve competing social or policy objectives, such as access to information, research, education and cultural development. Restrictions designed to achieve these objectives are generally known as "exceptions and limitations" to exclusive IPRs. These exceptions and limitations generally appear in two forms. The first form permits third parties to engage in specified uses of intellectual property products without the permission of the rights holder and without the payment of remuneration. The second form is known as a "compulsory licence." Compulsory licences allow third parties to use intellectual property products without the owner's permission, but only upon the payment of adequate compensation. To prevent both forms of exceptions and limitations from eviscerating IPRs altogether, intellectual property agreements impose specific constraints the ability of member states to adopt them.

Enforcement provisions: The grant of IPRs in national laws would be meaningless without adequate and effective mechanisms to enforce those rights. For this reason, recent intellectual property agreements specify the types of enforcement provisions that member states must adopt in their national laws. These provisions include the imposition of civil and criminal penalties against any person who engages in acts of exploitation reserved to the owner of an intellectual property product without the owner's authorization. The penalties include civil judicial proceedings for monetary damages or an injunction to prevent the continued unauthorized use of the product and criminal proceedings commenced by the government itself.

The "Minimum Standards" Framework of Multilateral IPR Agreements

Taken together, these core provisions of international IPR agreements impose significant legal obligations on member states. The agreements do not, however, purport to definitively address all of the issues raised by the grant of legal protection to intellectual property products. For this reason, the treaties are often referred to as "minimum standards" agreements in that they create only a basic floor of legal protection to which all member states must adhere.

There are three important consequences of this minimum standards framework. First, it allows member states the discretion to interpret and apply those provisions of the treaties that are ambiguous or that reasonably permit more than one construction. Second, a minimum standards approach permits, but does not require, states to grant additional IPR protections within their national laws. And third and perhaps most importantly, the framework leaves member states free to enact laws that serve other political, economic or social objectives, even where those objectives are in tension with IPRs, provided that those laws are not inconsistent with the terms of IPR agreements.

In the last five years, however, an increasing number of developing countries have entered into bilateral or regional trade and investment treaties with the United States or with the European Community. These treaties often contain higher standards of intellectual property protection than those specified in multilateral intellectual property agreements. (Drahos, 2001; GRAIN, 2001)

They thus restrict the minimum standards approach contained in the multilateral agreements, for example, by requiring states to enact specific intellectual property protection rules in their national laws or by requiring developing countries to follow the standards contained in multilateral agreements to which they are not parties.

The minimum standards framework (as limited by bilateral or regional treaties requiring higher standards of protection) provides a methodology for analysis but not a solution for every potential clash between IPRs and other governmental policies. This is particularly true for plant varieties and plant breeders' rights, an area of intellectual property protection regulated by several international IPR agreements and subject to diverse standards of legal protection under different domestic laws. To understand how national governments can reconcile competing and sometimes conflicting domestic laws and objectives consistently with the obligations imposed by different international IPR agreements, it is first necessary to examine other international agreements and international institutions relating to plant genetic resources that have promulgated policies in tension with IPRs.

Identifying the relevant international agreements and institutions relating to intellectual property rights in plant varieties.

Before turning to a more detailed discussion of policy objectives, it is important first to identify the principal international institutions and international agreements that generate legal rules and standards relating to IPRs in plant varieties and plant genetic resources, as well as the institutions and agreements that generate rules and standards in tension with IPRs. Although a comprehensive discussion of these institutions and agreements is beyond the scope of this study, a basic familiarity with their most important elements is necessary to understand the legal regime of intellectual property rights in plant genetic resources for food and agriculture.

World Intellectual Property Organization ("WIPO")

The World Intellectual Property Organization ("WIPO") is a specialized agency of the United Nations charged with "promot[ing] the protection of intellectual property throughout the world." (WIPO Convention, art. 3(i)). The WIPO Secretariat undertakes wide variety of activities relating to IPRs, including hosting diplomatic conferences of government representatives seeking to negotiate new international treaties. WIPO's staff also provide technical assistance and training to member states and their national intellectual property offices, especially in developing countries. More recently, WIPO has created standing, expert and intergovernmental committees that conduct studies on particular intellectual property topics and generate nonbinding guidelines and recommendations for consideration by WIPO members.

WIPO's recent activities in the area of plant genetic resources have been considerable. Issues relating to the intersection of IPRs and PGRs have been raised during the negotiation of two multilateral patent agreements (the Patent Law Treaty and the Substantive Patent Law Treaty) and have been a principal subject of discussion and study in the recently created Intergovernmental Committee on Intellectual Property and Genetic Resources, Traditional Knowledge and Folklore ("IGC"). Developing states first sought to raise issues relating to the intersection of IPRs and PGRs during the WIPO-sponsored negotiation of the Patent Law Treaty in 1999.

They proposed the addition of an article in the treaty requiring applicants for inventions derived from genetic resources to demonstrate that they had received permission to access those resources from the country of origin. Industrialized countries opposed the proposal, arguing that it addressed substantive law issues that were inappropriate for inclusion in a treaty largely devoted to procedural issues. As a compromise, the WIPO Secretariat proposed the creation of a new intergovernmental committee (the IGC) to study the intellectual property aspects of genetic resources and traditional knowledge.

During the IGC's first five sessions between April 2001 and July 2003, WIPO members have supported a wide-ranging work programme for the Committee, including: (1) creating a searchable database of clauses in contracts that regulate access to genetic resources or require benefit sharing; (2) studying technical issues raised by the disclosure of biodiversity-related information in patent applications; (3) creating databases of traditional knowledge; (4) identifying ways to document genetic resources and traditional knowledge in the public domain; and (5) studying the appropriate legal rules to protect traditional knowledge. Most recently, the WIPO General Assembly extended the IGC's mandate and authorized it to accelerate its work, which may include the development of new international agreements.

World Trade Organization ("WTO"): The WTO is a global intergovernmental organization dealing with the rules of trade between nations. It was established in 1994 at the conclusion of the Uruguay Round of trade negotiations held under the auspices of the General Agreement on Tariffs and Trade ("GATT"). As of July 2004, 147 states and customs territories ("WTO Members" or "Members") had joined the organization.

The WTO's substantive obligations are contained in a series of treaties relating to international trade, including agreements on trade in goods, services, agriculture, sanitary and phytosanitary measures, trade-related investment measures, technical barriers to trade, textiles and clothing and intellectual property rights. These treaties are linked together as annexes to the Agreement Establishing the WTO. Disputes between WTO Members relating to any of these treaties are to be resolved by the WTO Dispute Settlement Body, which is composed of *ad hoc* dispute settlement panels and a standing Appellate Body of trade experts. A WTO Member whose national laws or practices are challenged by another WTO Member and found to be incompatible with its WTO treaty obligations by the Dispute Settlement Body must modify those laws or practices or face the prospect of trade sanctions.

A more detailed discussion of the dispute settlement process in the context of the Agreement on Trade Related Aspects of Intellectual Property Rights ("TRIPs" or "TRIPs Agreement"), is provided in paragraph 2.3.1.3.3 below. TRIPs is the WTO agreement that addresses IPRs. It is discussed in detail in section 2.3 below.

Convention on Biological Diversity ("CBD") : The CBD was opened for signature in 1992 and entered into force in 1993. As of July 2004, 188 states had ratified this agreement.

The CBD's main objectives are the conservation of biological diversity, the sustainable use of its components, the fair and equitable sharing of benefits arising out of the utilization of genetic resources and the preservation of indigenous knowledge. (art. 1) The CBD also recognizes that nation states have the sovereign right to exploit their own resources and the authority to determine the conditions of access to them. (arts. 3 and 15)

One of the mechanisms by which the CBD achieves its objectives is *in situ* conversation of plant genetic resources. Conservation *in situ* involves the preservation of ecosystems and natural habitats and the maintenance of viable populations of species in those settings. Such conservation occurs, for example, where farmers and indigenous communities safeguard traditional plant varieties in the locations where they grow naturally or are cultivated.

Although the CBD does not expressly refer to any international IPR agreements, it contains numerous provisions relating to IPRs, principally in article 16. In particular, article 16(5) recognizes that IPRs "may have an influence on the implementation" of the CBD. The article obliges member states to cooperate in order to ensure that IPRs are "supportive of and do not run counter to" the treaty's objectives. Other provisions make clear that the CBD is to be interpreted so as to preserve the rights of IPR owners recognized in international law. For example, articles 16(2)-(4) state that the transfer of technology and measures taken to gain access to such technology shall be consistent with the adequate and effective protection of IPRs recognized in international law. Thus, for example, where a government encourages foreign direct investment in industrial technologies (such as a biotechnological process used to insert new genetic sequences into existing plant varieties), it must respect any patent rights that the owner of that technology has acquired to protect it.

Over time, the biodiversity regime's approach to intellectual property protection has evolved beyond the text of the CBD. The Conference of the Parties (COP) - the convocation of CBD members that determines how the Convention should be applied and implemented - has given detailed attention to harmonizing IPRs with the CBD's objectives. In particular, developing countries active in the COP, together with the support of nongovernmental organizations, have expressed concern about the adverse effects of IPRs and have sought to harness intellectual property rules to promote compliance with the Convention.

In response to this concern, official COP statements have stressed the need to "promot[e] increased mutual supportiveness and integration of biological diversity concerns and the protection of intellectual property rights." (CBD Decision IV/15, para. 9) Members of the COP have worked toward this objective by gathering information, commissioning case studies, holding workshops and drafting guidelines and recommendations.

For example, in April 2002, the COP adopted the "Bonn Guidelines on Access to Genetic Resources and Fair and Equitable Sharing of the Benefits Arising Out of their Utilization." (CBD Decision VI/24, App. II, Annex, Part C) The Guidelines' most important recommendation encourages applicants for IPRs such as patents or plant breeders rights to disclose the country of origin of the genetic resources or the traditional knowledge upon which those IPRs are based. The Guidelines advocate these disclosures to monitor whether applicants for IPRs have obtained the prior informed consent of the country of origin and complied with the conditions of access (if any) that that country has adopted. (Id.)

Although the COP members adopted the Bonn Guidelines by consensus, the Guidelines' reliance upon intellectual property laws to promote compliance with the obligations of the CBD remains controversial. For example, a 2003 Communication from the European Commission to the European Parliament on the EC's implementation of the Bonn Guidelines considers whether to adopt "a self-standing disclosure requirement" for patent and plant variety protection applicants. (EC Implementation of Bonn Guidelines, COM (2003) 821) If adopted, such a disclosure requirement would not, however, be incorporated into national or European intellectual property laws. To the contrary, the failure to comply with the new disclosure requirement would, according to the EC Communication, "only have consequences outside the field of patent law." However, the EC Communication also states that the EC and its member states "should be ready to discuss, in the relevant international fora, the possibility of introducing *under intellectual property law* the same disclosure requirement... as a formal condition for patentability and not only as a self-standing obligation."

***The Consultative Group on International Agricultural Research ("CGIAR")*:** CGIAR is an informal association of public and private donors founded in 1971 that supports an international network of agricultural research centres, each with its own governing body. CGIAR's mission is to engage in research, in partnership with other public and private entities, to promote sustainable agriculture in developing nations. With respect to plant genetic resources, CGIAR's principal method for achieving this objective is *ex situ* conversation. Networks of gene banks within CGIAR store and conserve seeds and propagating materials outside of their natural habitat for future use by farmers, researchers and breeders. The CGIAR network holds the world's largest *ex situ* collection of plant genetic resources for food and agriculture.

***International Plant Genetic Resources Institute ("IPGRI")*:** IPGRI is the world's largest international institute dedicated to the conservation and use of plant genetic resources. Founded in 1974, IPGRI focuses on conservation, management and preservation of the diversity of plant genetic resources, through domestic, regional and international programmes and research initiatives.

***The International Undertaking on Plant Genetic Resources (the "Undertaking")*:** FAO has helped to generate several nonbinding international instruments relating to plant genetic resources. The Undertaking, the first of these instruments, was adopted in 1983. As of 2000, 113 states were signatories to the Undertaking, thus pledging themselves to implement the recommendations it contains. For many years, the Undertaking served as the central legal instrument in FAO's global system for plant genetic resources, a system that includes a fund for the equitable sharing of benefits and a mechanism to give early warning about genetic resources under threat.

The Undertaking's principal objectives are to ensure that the need for conservation is globally recognized and that sufficient funds for this purpose are made available; to assist farmers and farming communities in the protection and conservation of PGRs and of the natural biosphere; and to allow farmers, their communities and countries to participate fully in the benefits derived from improved uses of PGRs, including through plant breeding. (WT/CTE/125, para. 11)

In its initial formulation, the Undertaking challenged a private property rights approach to plant genetic resources by declaring that all such resources, whether as cultivated by farmers in the field or modified through breeder innovations, were part of the "common heritage of mankind and consequently should be available without restriction." (art. 1). An interpretation issued adopted by FAO in 1989, however, clarified that plant breeders' rights were not incompatible with the Undertaking. It also recognized the interrelationship between the rights of traditional farmers (whose practice of saving seeds provided the raw genetic materials for innovation) and the rights of plant breeders (who use technology to achieve that innovation).

***The International Treaty on Plant Genetic Resources for Food and Agriculture ("ITPGR")*:** On 3 November 2001, an intergovernmental conference sponsored by FAO adopted the text of a legally binding international agreement on plant genetic resources. The ITPGR entered into force on 29 June 2004. As of July of that year, the treaty had been ratified by 55 states and signed by an additional 50 countries. The ITPGR not only codifies and updates the nonbinding principles set out in the Undertaking and its subsequent revisions, but also contains provisions relevant to IPRs in plant genetic resources and plant varieties. Because of its importance to the subject of this study, an extended discussion of the ITPGR appears in section 4.3 below.

Specific Policy Objectives in Tension with IPRs

Having identified the principal institutions and agreements relating to plant genetic resources, the next sections address specific critiques of IPRs as applied to plants and plant varieties and the policy arguments that inform those critiques.

***Preserving genetic diversity*:** Granting IPRs to plant breeders has uncertain consequences for preserving plant genetic diversity. Given the incentives that IPRs create for private parties to invest in research and breeding techniques relating to new plant varieties, it might be thought that IPRs would lead to an increase in plant genetic diversity over time. A number of commentators have argued, however, that diversity is eroded rather than enhanced by granting IPRs to plant breeders. According to this view, *in situ* conservation by indigenous farmers diminished as they began to rely on commercial plant breeders for seeds and other propagating material. Rather than using informal breeding techniques to experiment with the creation of new varieties suitable for local growing conditions, indigenous farmers came to depend upon third party plant breeders to provide them with seeds possessing uniform genetic characteristics. The plant varieties that have come to dominate agriculture as a result of this dependence may possess many beneficial characteristics, but they do not enjoy the adaptive abilities of less well known and informally bred varieties.

To evaluate these competing assertions about the effects of IPRs on plant genetic diversity, empirical testing would be useful. Unfortunately, empirical studies of the causal impact of IPRs on plant genetic diversity are difficult to devise because of the uncertainty of screening out other relevant causal variables. Empirical studies that do exist often focus on industrialized countries rather than developing states and tend to rely on anecdotal evidence. As a result, the broader applicability of their conclusions is uncertain and often controversial.

Farmers' rights: A second challenge to IPRs concerns the relationship between farmers' rights and IPRs in plant varieties. The concept of farmers' rights was developed to reflect the contributions that traditional farmers, particularly in the developing world, have made to the preservation and improvement of plant genetic resources. FAO Resolution 5/89 defines farmers' rights as "rights arising from the past, present and future contributions of farmers in conserving, improving and making available plant genetic resources, particularly those in centres of origin/diversity." Such rights are also recognized in article 9 of the ITPGR.

Farmers' rights are in tension with IPRs for plant breeders because many farmers and farming communities do not claim exclusive rights in the cultivated landraces and plant varieties they have cultivated over time. Moreover, the subject matter requirements of existing IPR laws, which as explained below are designed to protect innovations in new and clearly distinguishable plant *varieties*, often cannot accommodate contributions of individual farmers using more informal methods to select for better crops or sought-after plant *characteristics*.

Advocates of farmers' rights have developed different approaches to address this situation and to reward farmers for their contributions to plant genetic diversity. The first approach involves situating the traditional practices of farmers as exceptions to the exclusive rights of plant breeders under existing IPR laws. In other words, breeders are precluded from demanding payment from farmers who engage in certain farming practices, such as saving seeds and planting seeds saved from prior purchases, or informally exchanging seeds. A second approach seeks to modify existing IPR laws so as to permit farmers themselves to claim exclusive rights in the plant varieties they cultivate informally. A third approach involves recognizing farmers' rights not through IPRs but through benefit sharing mechanisms, such as financial payments and technology transfers, that compensate farmers for their contributions to plant genetic diversity. This last approach - which is recognized in article 9 of the ITPGR- questions whether farmers in fact have "rights" as that term is understood within an intellectual property paradigm, while acknowledging the need to reward their contributions to plant genetic diversity.

Traditional knowledge: An issue closely related to farmers' rights is the recognition and protection of the plant-related knowledge, innovations and practices of indigenous and local communities. Advocates assert that those claiming IPRs in plant genetic resources and plant varieties often utilize such knowledge without adequately acknowledging the contributions of the communities who possess it. Mechanisms to redress this problem are similar to those discussed above relating to farmers' rights, a linkage reflected in article 9 of the ITPGR. They include modifying existing IPR laws to recognize traditional knowledge as a form of intellectual property and protecting traditional knowledge through *sui generis* national laws that recognize the communal ownership of much of traditional knowledge. In the fall of 2000, the World Intellectual Property Organization (WIPO) established a new Intergovernmental Committee on Intellectual Property and Genetic Resources, Traditional Knowledge and Folklore to study these issues.

Regulating access to plant genetic resources: Plant breeders and others seeking to develop plant-related innovations need access to existing stocks of plant germplasm for breeding, research and development. Access issues arise for both *in situ* and *ex situ* collections of seeds and plant propagating material. With respect to *in situ* collections, a number of states have enacted national access laws and regulations to control third party access. These access controls are consistent with the CBD, which grants states the sovereign right to control their plant genetic resources. They also further the CBD's benefit sharing objectives by conditioning access upon the current or future payment of compensation or the transfer of technology to the state providing access. The specific conditions of access are often specified in Material Transfer Agreements ("MTAs") entered into between the entity (whether governmental or private) that owns or controls the germplasm and the entity (usually private) that seeks access to it.

Similar access issues arise with regard to *ex situ* collections of plant germplasm, including the seed banks maintained under the auspices of CGIAR. Although pursuant to a 1994 agreement between CGIAR and FAO the autonomous agricultural research centres under CGIAR's auspices each create their own access regulations, certain designated germplasm is to be held by the centres "in trust for the benefit of the international community." (art. 3(a)) With respect to IPRs, the 1994 agreement provides that a centre "shall not claim legal ownership over the designed germplasm, nor shall it seek any [IPRs] over that germplasm or related information." (art. 3(b)). The centres also commit to apply the same restrictions on IPRs in MTAs with third parties to whom the centre transfers designated germplasm. (art. 10)

"Biopiracy" and property rights in unimproved plant materials: Issues of access to plant genetic resources are linked to larger controversies concerning the propriety of granting IPRs in the raw materials accessed.

Under settled principles of intellectual property law, unimproved plant germplasm already in the public domain cannot be removed and privatized. Nevertheless, there are several reported cases and one empirical study which indicate that plant breeders and commercial entities have on occasion been granted IPRs in wild plant varieties or in germplasm found in CGIAR seed collections. It is unclear whether such rights were granted because of inadvertence, insufficient or inaccurate information provided by the IPR applicant or particularities in national laws that limit the sources to which granting authorities may refer in determining what materials are in the public domain. Where such rights have been granted, however, states and interested NGOs have often succeeded in petitioning the authorities in other nations to remove legal protections they had previously granted.

Attempts to claim IPRs in unimproved plant genetic resources have often been labelled as a form of "biopiracy." Biopiracy is not a legal term of art, however and has been loosely used to refer to any act by which a commercial entity seeks to obtain IPRs over biological resources, including plant varieties, that are seen as "belonging" to developing states or indigenous communities. Thus, even where an IPR claim relates to improvements to raw plant materials, certain governments and NGOs have labelled the entity seeking legal protection as a biopirate if it has not provided a fair return to those who granted access to the raw materials.

Plant breeders' research interests: Even as between groups of plant breeders, the scope of IPRs in plant varieties can be controversial. Tensions arise between first generation breeders who have secured legal protection for new varieties and second generation breeders who seek to utilize those new varieties to develop still more varieties. As with farmers' rights, it is possible to use the exceptions and limitations provisions of national IPR laws to permit second generation innovators to engage in such activities without the authorization of first generation breeders.

International IPR Agreements Regulating Plant Varieties and Plant Breeders' Rights

This section identifies and discusses the provisions of three different international IPR agreements that protect plant varieties and plant breeders' rights.

The two major treaty systems that regulate these issues are the agreements established under the auspices of the *Union internationale pour la protection des obtentions vegetales* ("UPOV"), and the TRIPs Agreement included within the family of treaties administered by the World Trade Organization ("WTO"). These two treaty systems each contain a comprehensive set of rules for their members regarding IPRs over plant varieties.

In short, the UPOV treaties adopt a *sui generis* system of protection (that is, a system that is unique, or of its own kind) especially tailored to the needs of plant breeders. The TRIPs Agreement requires WTO Members to protect new plant varieties using patent rights, a *sui generis* system or some combination thereof. Because TRIPs provides states with this flexibility and because the treaty has an uncertain relationship to the previously adopted UPOV conventions, national governments face a wide array of options in choosing the intellectual property regime applicable to plant varieties. This section of the study outlines the requirements imposed by these two treaty systems, and Part III of the study then identifies and analyses these options.

The UPOV Acts

The first UPOV Act was drafted in 1961, principally by industrialized governments seeking to provide protections for plant breeders in their own and overseas markets. The UPOV was later revised in Acts adopted in 1972, 1978 and 1991.

As of July 2004, 55 states were parties to UPOV - 29 to the 1978 Act, 24 to the 1991 Act and 2 to the 1972 Act. (States Parties to the UPOV, Status on 30 June 2004). As a result, this study will focus on the two most recent UPOV Acts. Many of the accessions to these Acts are quite recent, with a number of developing states and countries in transition to a market economy acceding in the 1990s. The 1991 Act entered into force on 24 April 1998 and on that same date the 1978 Act was closed to future accessions except by a few states already in the process of adhering to it.

As explained in section 1.3.1 above, countries generally give domestic effect to the UPOV Act to which they are a party in one of two ways. In "automatic incorporation" states, courts and administrative agencies directly apply and enforce the Act, although implementing legislation is often needed to authorize administrative agencies to process applications to protect new plant varieties.

In "legislative incorporation" states, by contrast, the UPOV Act does not become enforceable in domestic law until the state enacts a national plant variety protection law that conforms to the Act's requirements.

The 1978 UPOV Act

The 1978 UPOV Act adopts most of the international IPR obligations set out in Part I above, including a definition of the applicable subject matter and protected material, eligibility requirements, exclusive rights, national treatment, reciprocity, terms of protection and exceptions and limitations to exclusive rights. It does not, however, contain any provisions on MFN treatment or enforcement.

Subject Matter Requirements

Limited number of protected genera or species : Not all plant varieties must be protected under the 1978 Act. Rather, article 4 provides that member states are to progressively extend protection to an increasing number of genera or species, beginning with five on the date the treaty enters into force for that state and ending with twenty-four within eight years. In addition, member states are free to limit the Act's application within a particular genus or species to varieties with a particular manner of reproduction or multiplication, or a certain end-use.

Preclusion of dual protection with breeder's right and patent : The 1978 Act permits its signatories to protect plant varieties either with a distinct breeder's right or with a patent. However, article 2(1) precludes member states from granting both forms of protection "for one and the same botanical genus or species."

Protection of discovered varieties required : Although the 1978 Act focuses on plant varieties created through classical breeding methods, it is generally accepted that the treaty requires member states to protect varieties which have been discovered. This has been inferred from article 6.1(a) which indicates that a protected variety may result from a natural source of initial variation.

Eligibility Requirements

Assuming that a plant variety falls within a protected genera or species, it is eligible for protection under the 1978 Act only if it is: (1) new, (2) distinct from existing or commonly known varieties, (3) homogenous or uniform and (4) stable. (art. 6) When a variety fulfils these four criteria, it is listed in a national register or catalogue which publicly discloses that the variety is protected. (FIS/ASSINSEL, 2001a).

Novelty : To avoid protection for plant varieties that have already been exploited or are a matter of common knowledge, a new variety in which a breeder seeks protection must not have been sold on the market for more than a specified period of years prior to the date of application for protection. The 1978 Act specifies the maximum number of years during which such pre-application sales have occurred, with different periods of time set for different types of plants as well as for sales within the territory of the applicant state versus the territory of other states. (art. 6(1)(b)).

Distinctness : The 1978 Act states that a protectable plant variety must be "clearly distinguishable in one or more important characteristics from any other variety whose existence is a matter of common knowledge at the time when protection is applied for." (art. 6(1)(a)).

Although the treaty itself does not further define distinctness, the Guidelines for the Conduct of Tests for Distinctness, Homogeneity and Stability (UPOV Guidelines) use both qualitative and quantitative plant characteristics, including such visible attributes as leaf shape, stem length and colour, to determine if the difference between varieties is "clear and consistent." As explained below in the analysis of the UPOV 1991 Act, the concept of distinctness is critical to determining the scope of a breeder's right in plants that are closely related but not identical to a protected variety.

Homogeneity : Under the 1978 UPOV Act, a variety has to be "sufficiently homogeneous, having regard to the particular features of its sexual reproduction or vegetative propagation." (art. 6(1)(c)). The UPOV Guidelines further clarify that to be considered homogeneous, the variation shown by a variety must be "as limited as necessary to permit accurate description and assessment of distinctness and to ensure stability." The homogeneity requirement has been criticized by commentators as discouraging variability in plant varieties that are often useful for sound agricultural practices and as denying protection to breeders of cultivated landraces that exhibit diversity traits. It is thus one of the sources for the criticisms identified in paragraph 1.3.6.1 above that plant breeders' rights are reducing plant genetic diversity by rewarding breeders of uniform plant varieties.

Stability : The stability requirement is a temporal one, requiring the breeder to show that the essential characteristics of its variety are homogeneous or uniform over time, even after repeated reproduction or propagation. (art. 6(1)(d)) In practice, what has been shown to be homogeneous is usually considered to be stable as well. For this reason, the stability requirement has engendered the same sort of critiques as the uniformity requirement in its preclusion of protection for cultivated landraces and other traditional plant varieties.

Protected material : The 1978 Act requires its signatories to protect a variety's reproductive or vegetative propagating material. The Act does not require protection of harvested material, with the exception of ornamental plants that are used for commercial propagating purposes. (art. 5(1))

Breeders' exclusive rights : Under article 5 of the 1978 Act, any person seeking to engage in the following three acts, with respect to a protected variety's reproductive or vegetative propagating material, must obtain the prior authorization of the breeder: (1) production for purposes of commercial marketing, (2) the offering for sale and (3) marketing. The 1978 Act does not, however, require member states to extend these exclusive rights to harvested material or other marketed products.

National treatment and reciprocity : Member states must grant the three exclusive rights referred to in the preceding paragraph in the same manner to both national breeders and to breeders who reside in or are nationals of other 1978 Act member states. However, where a state extends legal protection to a specific genus or species, or where it provides more extensive exclusive rights to breeders than the rights required under the treaty, reciprocity is permitted. Thus, a state providing these additional rights may restrict protection to breeders from those member states that apply the Act to the same genus or species, or that provide such additional exclusive rights to their own nationals. (arts. 3 and 5(4))

Term of protection : The 1978 Act requires a minimum term of protection of fifteen years, with the exception of vines, forest trees, fruit trees and ornamental trees, which are protected for no less than eighteen years.

Exceptions and limitations : Two major exceptions and limitations to exclusive rights exist under the 1978 Act: (1) a breeders' exemption and (2) a farmers' privilege. The Act also permits members to impose compulsory licences.

Breeders' exemption: This exemption in article 5(3) precludes member states from granting to breeders of protected varieties the right to authorize or refrain from authorizing other breeders seeking to use the protected variety to create new varieties or to market those new varieties. States are permitted to grant breeders such an authorization right only if the repeated use of the protected variety is necessary for the commercial production of the new variety. According to the International Association of Plant Breeders and the International Seed Federation, this breeders' exemption "is essential for continued progress from plant breeding." (FIS/ASSINSEL, 1999)

Farmers' privilege : The focus of the 1978 Act on commercial exploitation of protected plant varieties has been interpreted to allow the use of seeds and propagating material for noncommercial purposes without the breeder's prior authorization. In national plant variety protection laws, this implicit noncommercial exception most frequently benefits farmers who purchase the seeds of protected varieties. The scope of this so-called farmers' privilege varies widely, however. Some nations only permit farmers to plant seeds saved from prior purchases to be used on their own land holdings, while others allow them not only to replant but also to sell limited quantities of seeds for reproductive purposes, a practice often referred to as "brown bagging."

Compulsory licences in the public interest : Article 9 of the 1978 Act permits members to restrict breeders' exclusive rights for "reasons of public interest." Where such restrictions are enacted to ensure the widespread distribution of the variety (such as where the breeder fails to supply the demand for the variety in a reasonable quantity and price or unreasonably refuses to license the variety to third parties), the breeder must receive equitable remuneration.

The 1991 UPOV Act

The limited scope of the 1978 Act led a number of member states of the UPOV to adopt a revised Act with enhanced rights for plant breeders. The major revisions of the 1991 Act are discussed below:

Subject Matter Requirements

Phased-in protection of all genera or species : Unlike its predecessor, the 1991 Act requires states to protect at least fifteen plant genera or species upon ratifying or acceding to the Act, and to extend protection to all plant varieties within ten years. (art. 3(2)) It also contains a definition of a plant "variety" as a "plant grouping within a single botanical taxon of the lowest known rank" which can be "defined by the expression of the characteristics resulting from a given genotype or combination of genotypes; distinguished from any other plant grouping by the expression of at least one of the said characteristics; and considered as a unit with regard to its suitability for being propagated unchanged." (art. 1(vi)) No definition of "variety" appears in the 1978 Act, which indicates that member states to that earlier treaty have greater discretion in defining the characteristics of plant groupings that qualify for protection.

Dual protection with breeders' right and patent permitted : In response to demands from breeders in industrialized counties, the 1991 Act removed the 1978 Act's ban on dual protection and now permits member states to protect the same plant variety with both a breeders' right and a patent.

Protection of discovered varieties : The 1991 Act makes explicit the 1978 Act's implicit requirement that discovered varieties be protected. It does so through article 1(iv)'s definition of a "breeder" as including a "person who bred, or discovered and developed, a variety."

Eligibility Requirements

The four eligibility requirements that must be demonstrated to merit protection for a specific variety - novelty, distinctness, uniformity and stability - are preserved in the 1991 Act, subject to only minor changes in scope and wording. (arts. 7-9) Thus, the 1991 Act has received the same criticism as the 1978 Act for its encouraging of genetic standardization and its inability to protect more diverse plant varieties, traditional varieties or cultivated landraces.

Breeders' Exclusive Rights in Protected Material

Extensive additions to the 1978 Act were made with respect to the exclusive rights enjoyed by breeders in protected material of plant varieties.

Enumerated exclusive rights in propagating material : The breeder's prior authorization must be obtained for the use of reproductive or vegetative propagating material of the variety for (1) production or reproduction, (2) conditioning for the purpose of propagation, (3) offering for sale, (4) selling or marketing, (5) exporting, (6) importing and (7) stocking for any of these purposes.

Extension of rights to harvested material and products : These exclusive rights apply not only to propagating material but also to harvested material, where the harvest has been obtained through an unauthorized use of the propagating material and the breeder has not had a reasonable opportunity to exercise his or her right in relation to that material. In addition, member states are permitted but not required to provide additional rights to breeders, including rights in products made directly from harvested material.

National Treatment Required

Whatever the particular exclusive rights member states adopt in their plant variety protection laws, those rights must also be provided to the nationals of other 1991 Act member states. Unlike the 1978 Act, granting rights only on condition of reciprocity is not permitted. (art. 4) For example, if a 1991 Act member chooses to grant more expansive exclusive rights to breeders than the exclusive rights required by the 1991 Act, it must grant those rights to breeders from all other 1991 Act member states. The state cannot choose to grant such expansive rights to only those 1991 Act states that also provide the same level of rights protection to the first state's nationals.

Terms of Protection

The 1991 Act extends the term of protection to 20 years, and requires a 25-year term for tree and vine varieties.

Exceptions and Limitations

A second area of major revision concerns the reduced scope of exceptions and limitations to breeders' exclusive rights, which is found in article 15 of the 1991 Act.

Private, noncommercial exception : The 1991 Act makes explicit what was only implicit in the 1978 Act, namely that private, noncommercial activities with respect to new varieties are outside of the breeder's control. This exception would presumably permit subsistence farmers to use protected seeds and other propagating material for their own consumption.

Research exception : The 1991 Act also recognizes that breeders cannot restrict "acts done for experimental purposes." This exception would permit research and testing of protected varieties for scientific purposes that does not lead to commercial exploitation.

Limited breeders' exemption : Like its predecessor, the 1991 Act recognizes the right of breeders to use protected varieties to create new varieties. However, this exception is itself restricted in its application to such new varieties as are not "essentially derived" from protected varieties. (arts. 14(5) and 15) The drafters added this restriction to prevent second generation breeders from making merely cosmetic changes to existing varieties in order to claim protection for a new variety. The concept of essential derivation has proved highly controversial in practice, however. Breeders have been unable to agree on a definition of the minimum genetic distance required for second generation varieties to be treated as not essentially derived from an earlier variety and thus outside of the first breeder's control. (FIS/ASSINSEL, 1999). Nevertheless, the overall effect of this provision of the 1991 Act has been to narrow the breeders' exemption and expand the IPRs of first generation breeders.

Limited farmers' privilege : The 1991 Act recognizes an explicit farmers' privilege, but one that is limited in scope. In particular, each member state may enact a provision in its national plant variety protection laws permitting farmers to use for propagating purposes "on their own holdings" the product of the harvest which they obtained by planting a protected variety "on their own holdings." In addition, this privilege must be exercised "within reasonable limits and subject to the safeguarding of the legitimate interests of the breeder." (art. 15(2)) Several points can be inferred from this language.

First, unlike the 1978 Act, the 1991 version of the farmers' privilege does not authorize farmers to sell or exchange seeds with other farmers for propagating purposes, a limitation that commentators have criticized as inconsistent with the practices of farmers in many developing nations, where seeds are exchanged for purposes of crop and variety rotation.

Second, according to the International Association of Plant Breeders, the "reasonable limits" language in article 15(2) requires states to restrict the acreage, quantity of seed and species subject to the farmers' privilege, while the requirement to safeguard breeders' "legitimate interests" requires farmers to pay some form of remuneration to the breeder for their privileged acts. (FIS/ASSINSEL, 2001a) The latter assertion is controversial, however, and has not been enacted in the national laws of all the 1991 Act member states. For example, the EC's Council Regulation on Community Plant Variety Rights, No. 2100/94 of 27 July 1994, requires such payments except in the case of small farmers, whereas the United States' Plant Varieties Protection Act of 1994, 7 U.S.C. Sec. 2543, does not.

Compulsory licences in the public interest : Article 17 of the 1991 Act contains a compulsory licence provision similar to that found in the 1978 Act. It permits members to restrict breeders' exclusive rights only for reasons of public interest and requires payment of equitable remuneration to the breeder whose rights are limited.

Resistance to the 1991 Act

Many developing nations, particularly those in Africa, have resisted ratifying the 1991 Act or adopting it as the standard for their plant variety protection laws. The foreign ministers of the more than 50-member Organization for African Unity (now the African Union) issued a statement at a January 1999 meeting calling for a hold on IPR protection for plant varieties until an Africa-wide system has been developed that grants greater recognition to the cultivation practices of indigenous communities. However, at a subsequent meeting of the Organisation Africaine de la propriete Intellectuelle (OAPIO), patent officials from sixteen francophone African nations recommended that their countries adopt the 1991 Act. (Machipisa, 1999). As of the date of this study, Tunisia, Kenya and South Africa are the only African UPOV member states. Tunisia ratified the 1991 Act in August 2003, whereas the other two countries are parties to the 1978 Act. Other developing states parties to the 1978 Act are Bolivia, China, Ecuador, Kyrgyzstan, Mexico, Nicaragua, Panama, Paraguay, Trinidad and Tobago and Uruguay. These and other developing countries are, however, under pressure to ratify the 1991 Act as part of bilateral or regional trade and investment agreements.

The TRIPs Agreement

Although the UPOV Acts have provided IPR protection for plant varieties for more than forty years, their significance has recently been overshadowed by a different intellectual property treaty, the Agreement on Trade-Related Aspects of Intellectual Property Rights ("TRIPs" or the "TRIPs Agreement"). Adopted in 1994 as a treaty administered by the WTO, TRIPs is the first and only IPR treaty that seeks to establish universal, minimum standards of protection across the major fields of intellectual property, including patents, copyrights, trademarks, industrial designs, integrated circuits and trade secrets. Although the TRIPs Agreement devotes only minimal attention to plant breeders' rights or plant variety protection and does not even mention the UPOV Acts, its adoption has done more to encourage the legal protection of plant varieties than any other international agreement.

TRIPs as a Spur to Plant Variety Protection

TRIPs' influence on plant variety protection stems from the following sources: (1) its link to other international trade agreements; (2) its widespread adherence by states in both the industrialized and developing world; (3) its novel enforcement, review and dispute settlement provisions; (4) the requirement in TRIPs article 27.3(b) that its signatories must provide protection for plant varieties "either by patents or by an effective *sui generis* system or by any combination thereof"; and (5) a formal review of article 27.3(b) which was scheduled to be held in 1999. The following paragraphs briefly address each of these five issues. Subsequent sections devote more detailed treatment to the protection of plant varieties with patents and to the elements necessary to create an "effective *sui generis* system."

The Relationship of TRIPs to the WTO and the International Trading System

Unlike all prior intellectual property treaties, TRIPs is not a free-standing agreement concerned solely with IPRs. Rather, TRIPs is linked to a larger family of trade-related agreements concerning subjects such as trade in goods and services, agriculture, textiles and health-related restrictions on imports. All of these agreements were adopted within the WTO during the Uruguay Round of trade negotiations held between 1988 and 1994. As such, TRIPs was part of a global "package deal." Industrialized nations secured a commitment from developing nations to provide minimum standards of effective legal protection to intellectual property products, and in exchange developing nations received a commitment from industrialized countries to open their domestic markets to goods and other products manufactured in the developing world.

Widespread Adherence to TRIPs

The result of this global bargain was widespread adherence to all WTO Agreements, including the TRIPs Agreement. As of July 2004, 147 states or customs territories were obligated to comply with TRIPs by virtue of their membership in the WTO. In addition, many states (particularly those in the developing world) who became bound by TRIPs had not ratified earlier intellectual property treaties and were thus required to make significant changes to their national laws to bring them into compliance with the Agreement. In acknowledgement of this fact, TRIPs contains a phase-in provision pursuant to which most of the treaty's substantive obligations became binding for developed nations in 1996, but did not become binding on developing nations and nations in transition to market economies until 2000. Least developed nations are given the most leeway and are not required to implement the treaty's substantive obligations until 2006.

The Enforcement, Review and Dispute Settlement Provisions of TRIPs

In addition to its widespread adherence, the influence of the TRIPs Agreement can be traced to its unique provisions relating to the enforcement of IPRs within national laws, the review of those national laws by the TRIPs Council and the mechanism for settlement of disputes between states leading to rulings backed up by the threat of trade sanctions.

Enforcement provisions : Unlike prior IPR agreements, TRIPs does not only specify the minimum substantive requirements for various forms of intellectual property. In addition, it requires WTO Members to adopt "effective" provisions within their national laws to permit the owners of intellectual property products to enforce their rights against those who infringe them. (art. 41.1) These enforcement provisions include detailed judicial and administrative remedies, border measures and criminal procedures. (articles 41 to 61) To take just one example applied to plant varieties, a breeder whose new variety is sold commercially without permission must be able to bring a civil judicial action seeking an injunction to stop the conduct of the unauthorized seller and to recover damages from him or her.

Review provisions : Article 68 establishes a Council for TRIPs to "monitor the operation of th[e] Agreement and... Members' compliance with their obligations" under the treaty. Since its founding, one of the Council's principal functions has been to formally review the national laws, regulations and judicial decisions of WTO Members in each area of intellectual property covered by TRIPs.

Several goals are served by the TRIPs Council's review functions. First, the reviews create an incentive for governments to bring their national laws and practices into compliance with the Agreement so that they may present positive information to the Council. Second, the reviews identify for both the reviewed state and for other WTO Members areas of the law which may not be in full compliance with the treaty. And finally, TRIPs Council reviews provide a critical opportunity to publicize national laws and practices relating to IPRs which might otherwise be difficult to obtain.

In the area of plant varieties, the TRIPs Council has gathered and organized a considerable amount of useful information concerning national government practices. In December 1998 the Council prepared a detailed list of questions concerning plant variety protection. It asked WTO Members who were already obligated to protect plant varieties to respond to these questions by indicating the manner in which their national laws provide such protection, with other WTO Members requested to use their best efforts to furnish such information. The Council later received and published responses from 17 states and from the European Communities and its member states. (WTO Job No. 2689 of 7 May 1999). In 2001 and 2002, the Council reissued these questions and invited WTO Members that had not already done so to provide information on their plant variety protection practices. Additional responses were received from six states. Information distilled from some of these responses is discussed in section 3.3 below.

Dispute settlement provisions : Another significant innovation of TRIPs is its dispute settlement system. Although prior intellectual property agreements, including the UPOV Acts, contained provisions for filing of complaints against non-complying treaty parties before the International Court of Justice, no state ever invoked this dispute settlement option. It was widely believed that pursuing such a complaint would be perceived as an unfriendly act, would be time-consuming and costly and that states were unlikely to implement decisions of the court. The TRIPs Agreement removes each of these concerns by linking to a WTO Agreement known as the Dispute Settlement Understanding which contains a streamlined dispute settlement mechanism that is one of the most widely used and effective interstate dispute resolution systems in international law. Indeed, the mere existence of such a system creates strong incentives for WTO Members to bring their national laws into compliance with the WTO Agreements, including TRIPs, to avoid the possibility of dispute settlement proceedings. These strong incentives notwithstanding, some countries may not comply.

In such a case, where one WTO Member believes that another Member has failed to fulfill its obligations under the Agreement, it may initiate consultations under the Dispute Settlement Understanding with a view to resolving the dispute. If the parties fail to reach agreement, the complaining state may then request the WTO to convene a three-member panel of experts to review its allegations. Such panels generally issue a decision within six months. The panel's decision is adopted by the WTO's Dispute Settlement Body unless the losing party or parties elect to appeal the decision to the seven-member Appellate Body, a standing tribunal of seven trade experts who are authorized to review the panel's findings and issue a decision within three months. If either the Appellate Body or an unreviewed panel decision concludes that a WTO Member has violated the TRIPs Agreement, the Dispute Settlement Body will recommend that the defending state bring its national laws into compliance with the treaty. If the state fails to do so, the complaining state may then commence an arbitral proceeding to specify the amount of compensation that the defending state must pay to remedy the violation, or in lieu of such compensation the complaining state may seek authorization to impose trade sanctions on the non-complying Member.

As of July 2004, WTO Members (mostly developed nations) had commenced 24 dispute settlement proceedings against both developed and developing nations concerning a wide variety of intellectual property issues. (WTO Dispute Settlement: Index of Disputes Issues: 15 July 2004) Approximately one third of these disputes were resolved prior to a decision by a panel, with the defending state agreeing to modify its laws to bring them into compliance with the Agreement. In other cases, however, WTO Members complied only after a panel or the Appellate Body had issued a decision against them. And in a few instances, Members that have not modified their laws instead have negotiated settlements involving the payment of compensation to the complaining Member.

As of the date of this study, there have been no TRIPs dispute settlement proceedings relating to intellectual property protection for plant varieties. Whether such proceedings will be brought in the future depends on a number of variables, many of which are still uncertain.

Three interrelated reasons suggest that a dispute settlement proceeding concerning plant variety protection is a likely possibility: first, the significant difference of views among WTO Members over the scope of legal protection to be provided to plant varieties; second, the fact that developing country WTO Members first became obligated to protect plant varieties only in 2000; and third, the absence of plant variety protection in many developing countries' national laws prior to the 2000 deadline. Taken together, these variables suggest that the plant breeders of a developed WTO Member who learn that their protected varieties are being exploited without authorization in developing WTO Members due to inadequate national laws will pressure their governments to file a complaint against such states to compel them to adhere to their treaty obligations.

Other factors, however, suggest that a complaint over plant varieties is unlikely to be filed, particularly within the next five to ten years. First, the initiation of the Doha Round of trade negotiations in November 2001 (discussed in Part IV below) is likely to cause WTO Members to show restraint in their dispute settlement strategies while the multi-year negotiations are proceeding. This is especially so given that the Doha Round will likely provide WTO Members with the first meaningful opportunity to harmonize other international obligations (such as those set forth in the CBD and the ITPGR) with the plant variety and other IPR protection requirements of the TRIPs Agreement. Second, past practice has shown that WTO Members are most likely to file dispute settlement complaints where they are pressured to do so by the owners of intellectual property products. It is unclear whether plant breeders can make a sufficient showing that their overseas markets are being harmed by other states' failure to provide adequate legal protection for plant varieties. Third, WTO Members are generally reticent about filing complaints where their probability of success is unclear. Therefore, given the uncertain scope of protection for plant varieties currently required by TRIPs, if Members enact *some form* of plant variety protection in their national laws, other Members are unlikely to challenge those laws unless they can identify a clear violation of the Agreement. Nevertheless, the potential for dispute settlement proceedings (and the trade sanctions they can engender) will continue to provide strong incentives for states to enact plant variety protection laws.

The Unique Scope of Plant Variety Protection Required by TRIPs

Article 27.3(b) contains the only textual provisions of the TRIPs Agreement relating to plant variety protection. The article states in relevant part:

Members may also exclude from patentability: (b) *plants* and animals other than microorganisms; and essentially biological processes for the production of plants or animals other than non-biological and microbiological processes. However, *Members shall provide for the protection for plant varieties either by patents or by an effective* sui generis *system or by any combination thereof*. (emphasis added)

The meaning of this article has been the subject of significant debate among both WTO Members and nongovernmental organizations (NGOs) with differing views over the propriety of IPR protection for plant varieties. A detailed review of that debate and an analysis of article 27.3(b) is provided below. For present purposes, two overarching points are worth noting.

First, TRIPs' provisions on plant varieties do not refer to or incorporate any preexisting intellectual property agreements, including the 1978 and 1991 UPOV Acts. This omission contrasts sharply with other fields of intellectual property, such as patents, copyrights and trademarks, for which TRIPs expressly requires WTO Members to comply with the standards of protection contained in preexisting IPR agreements, such as the Berne Convention for the Protection of Literary and Artistic Works and the Paris Convention for the Protection of Industrial Property.

As a result of this omission, WTO Members are neither required to become members of UPOV nor to enact national laws consistent with either UPOV Act in order to comply with their obligations under TRIPs. Although the drafting history of TRIPs does not explain this markedly different treatment of plant varieties, it seems likely that compliance with UPOV was not required because so few WTO Members were party to UPOV and those who were could not agree upon which of its two most recent Acts should serve as the standard for protection.

Second, article 27.3(b) permits WTO Members to protect plant varieties using one of three distinct approaches: (1) patent law, (2) an effective *sui generis* system or (3) a combination of elements from both systems. Thus, unlike most other areas of intellectual property protected by TRIPs, article 27.3(b) expressly grants Members significant discretion to choose the manner in which they will protect plant varieties and it contemplates that that discretion may be exercised differently by different states.

This discretion and the opportunity for divergent outcomes it engenders has significant consequences. On the one hand, TRIPs' failure to incorporate and build upon the preexisting UPOV Acts may have "a deharmonizing effect," (Correa 1994a) with states within the UPOV system enacting one type of plant variety protection law and states outside of that system enacting a different kind of law (which may or may not resemble each other).

This could create significant complexities and uncertainties for plant breeders seeking to market protected varieties in different jurisdictions. On the other hand, this sanctioned diversity of legal approaches allows WTO Members to balance the protection of plant breeders' rights against the other important and competing societal goals identified in Part I, many of which are found in other international agreements. Seen from this perspective, article 27.3(b) provides a much needed "safe space" for governments to harmonize conflicting norms and policies - a space that is lacking in other areas of the TRIPs Agreement.

The Review of TRIPs Article 27.3(b)

The last sentence of article 27.3(b) states that "[t]he provisions of this subparagraph shall be reviewed four years after the date of entry into force of the WTO Agreement." Inasmuch as the WTO Agreement entered into force on 1 January 1995, the review contemplated by this article should have been conducted in 1999. The review was commenced but was not concluded, largely as a result of disputes between industrialized and developing nations over the scope of the review process. Governments submitted additional information and proposals to the TRIPs Council in 2000 and 2001, but no formal action was taken. With the launching of the Doha Round of trade talks in November 2001, however, the review of article 27.3(b) recommenced in earnest.

In June 2002, eleven developing states submitted a detailed proposal to amend TRIPs to "prevent systematic conflicts with the CBD arising from the implementation of TRIPS." (WTO Doc. IP/C/W/356, para. 11) The proposal seeks to compel all WTO Members to require applicants for patents relating to biological materials and traditional knowledge to disclose certain information as a condition of obtaining legal protection. (Id., para. 10)

The European Communities and Switzerland responded to the developing states' proposal with a compromise. The EU response calls for the negotiation of "a self-standing disclosure requirement," which, while not functioning as a new eligibility criterion for patent protection, "would allow Members to keep track, at [the] global level, of all patent applications with regard to genetic resources for which they have granted access." In June 2003, developing countries rejected this compromise and "reaffirm[ed] and strengthen[ed] their demand for a strong disclosure of origin mechanism" within TRIPs that would require "not only detailed information about who provided the [genetic] materials or the [traditional] knowledge used [in patent applications], but also positive proof of benefit sharing and of prior informed consent." African countries also proposed a new "Decision on Traditional Knowledge" to be incorporated into TRIPs.

"TRIPs Plus" Bilateral and Regional Agreements

In the decade since TRIPs was first adopted, the United States and the European Community have negotiated a growing number of bilateral and regional trade and investment treaties with developing countries. Many of these treaties contain provisions concerning IPRs. Commentators have referred to these treaties by the appellation "TRIPs plus" because they (1) contain intellectual property protection standards more stringent than those found in TRIPs, (2) obligate developing countries to implement TRIPs before the end of its specified transition periods or (3) require such countries to accede to or conform to the requirements of other multilateral intellectual property agreements. As one study recently noted, "[t]o the extent that ["TRIPs plus" treaties] include these additional aspects, they are pushing harmonization forward at a pace that is greater than is apparently possible within the framework of the WTO." (OECD, 2003) In addition, once a developing state enacts this higher level of IPR protection in its national laws, TRIPs' MFN clause obligates that state to extend the same protection to *all other WTO Members*. In this way, bilateral agreements can be used to ratchet up the level of protection for IPRs beyond what is required in the TRIPs Agreement.

Several "TRIPs plus" treaties contain provisions relating to plant variety protection that go beyond the minimum level of intellectual property protection specified in TRIPs.

For example, the recently concluded U.S.-Central America Free Trade Agreement (applicable to Costa Rica, El Salvador, Guatemala, Honduras, Nicaragua and possibly to the Dominican Republic), the 2002 U.S.-Chile free trade agreement, the 2000 U.S.-Jordan free trade agreement and the 2000 EU-Mexico free trade agreement, as well as certain Euro-Mediterranean Association agreements and the draft Free Trade Agreement of the Americas, all mandate UPOV as the appropriate mechanism to protect plant breeders' rights. (GRAIN, 2004b) The treaties also require these countries to ratify the 1991 UPOV Act within specified time frames. Such obligations limit the discretion enjoyed by these countries to tailor their national laws to achieve policy objectives in tension with IPRs.

Patent Protection for Plant Varieties Under TRIPs

The TRIPs Agreement mandates its signatories to provide patent protection for any inventions in all fields of technology, provided that the inventions are "new, involve an inventive step and are capable of industrial application." (art. 27(1)) However, with regard to plant-related inventions, TRIPs permits Members to exclude from patentability altogether "plants," "essentially biological processes for the production of plants" and "plant varieties." (art. 27.3(b)) Thus, as presently written, the TRIPs Agreement would permit WTO Members to decline to protect plant varieties with a patent (provided, as discussed in section 2.3.3 below, that they protect such varieties with an effective *sui generis* right).

Applicability of Patent Protection for Plant Varieties

For this reason, it may initially appear that governments can ignore patents as they consider how to protect plant-related innovations in their national legal systems. Nevertheless, a basic understanding of the patent provisions of the TRIPs Agreement and a comparison of those provisions with the plant breeders' rights approach contained in the UPOV Acts is essential for a number of reasons.

First, extending patent protection to plant-related inventions and innovations remains an option for national governments. Recall that TRIPs' "minimum standards" framework expressly contemplates that WTO Members may provide greater protection for IPRs than are mandated by the Agreement. Article 27.3(b) in particular invites Members to protect plant varieties with patents or with a combination of patents and a *sui generis* system. A number of governments in the industrialized world, including the United States, Japan, Australia, New Zealand, Sweden and the United Kingdom, have capitalized on this opportunity by permitting plant breeders to obtain patent protection in new varieties provided that the eligibility requirements for a patent have been met.

Second, because TRIPs does not require any patent protection for plant-related innovations, it follows as a matter of course that the treaty does not compel WTO Members to adopt any particular form of patent protection. This allows governments the option of including plant varieties within their existing utility patent statutes and/or of enacting a separate statute applicable exclusively to plants. (Utility patents are generally granted to "any new and useful process, machine, manufacture or composition of matter, or any new and useful improvement thereof... " In making this decision, however, it will be useful for governments to understand the traditional elements of patent law and the ways in which they may be varied within the confines of the TRIPs Agreement. The United States, for example, has enacted multiple plant variety protection laws. It was one of the first nations to adopt, in the Plant Patent Act of 1930, a unique form of protection applicable to asexually reproducing plants. In addition, since the mid 1980s, U.S. courts and the U.S. Patent and Trademark Office have concluded that plant breeders may also seek standard utility patents under the U.S. Patent Act. In December 2001, the U.S. Supreme Court confirmed that breeders could apply for both forms of protection with respect to the same variety. Such a result is prohibited by the 1978 UPOV Act but permitted under the 1991 Act to which the U.S. is a party.

Third, an understanding of patent principles is also necessary because of the difficulty in some nations of distinguishing between plant varieties (which are often excluded from national or regional patent laws) and other plant-related innovations (to which patent protection may be extended). In Europe, for example, article 53(b) of the European Patent Convention prohibits the patenting of "plant varieties" as such. Nevertheless, the European Patent Office recently confirmed that claims to patent protection that are broadly drawn to encompass "plants" or an invention broader than a single variety *may* be patented, even though such claims may encompass multiple varieties. As a result, it may be possible for plant breeders in Europe to fashion their claims to receive *de facto* patent protection for new plant varieties.

There are significant differences in approach between plant breeders' rights and patents. In the case of plant breeders' rights, the eligibility requirements for protection are not onerous, but the scope of protection granted is quite narrow, both in terms of exclusive rights and the various exceptions and limitations to those rights. Patent laws strike a very different balance. Eligibility requirements are high and difficult to meet, but once granted a patent conveys broad rights to exclude third parties from exploiting the patented invention. Depending on the needs and level of development of plant breeder industries within its territory, a government may decide that either or both forms of protection will provide the appropriate incentives to encourage plant-related research and innovation.

Fifth, and finally, a review of the TRIPs Agreement's plant-related patent rules will soon occur in the context of the Doha Round of trade negotiations. Such a review may produce an agreement to amend TRIPs to require WTO Members to extend patent protection to plant-related innovations or plant varieties. National governments which have thus far eschewed patent protection would then need to revise their national laws to make such protection available, in addition to or instead of the legal protection provided under national plant variety protection laws. At a minimum, therefore, governments interested in promoting plant-related innovations should be apprised of the relevant legal issues so that they may contribute in an informed way to the upcoming discussions.

Subject Matter Eligibility Requirements

According to the text of the TRIPs Agreement and authoritative commentary interpreting it, plant-related innovations may be patented if they are "inventions" that are novel, non-obvious and useful. Although commentators frequently assert that these cumulative requirements impose significant barriers to the patenting of plant varieties, in practice breeders have succeeded in patenting not only new varieties, but also hybrid plants and inbred and hybrid plant lines. As two commentators recently noted, since 1985 the U.S. Patent and Trademark Office "has granted hundreds of utility patents on all aspects of innovation relating to plant science: plants themselves, seeds, breeding methods and plant biotechnology."

Inventions : The TRIPs Agreement does not define the term "invention," thus leaving its precise meaning to national patent laws or regional patent rules such as those operating within the European Union.

All such laws recognize that an invention must be more than a mere discovery of a natural phenomenon or naturally occurring substance. Thus, for example, a breeder who merely identified the existence of a previously unknown wild variety or a cultivated landrace that was known only to an isolated indigenous community should not be permitted to patent that variety or landrace. The trend among the industrialized countries, however, and in the United States, Japan and the countries of the European Union in particular, is to recognize that an isolated and purified form of a natural substance may be patented. For example, article 3.2 of the 1998 Directive on the Legal Protection of Biotechnological Inventions states that "biological material which is isolated from its natural environment or produced by means of a technical process" may be patentable. (Directive 98/44/EC of 6 July 1998, O.J. L213) Perhaps because this approach involves making a very fine distinction between discoveries and inventions, many developing countries have declined to follow it, instead choosing to deny patent protection to plant materials found in nature even if they have been isolated or purified by human intervention. According to one commentary, this patent ban is compatible with the TRIPs Agreement provided that it does not extend to plants with modified or artificial plant gene sequences, which often significantly vary from naturally occurring substances and thus are properly classified as inventions.

A second attribute of a patentable invention is that an applicant seeking such a patent must disclose the invention in a sufficiently clear and complete manner that permits a person skilled in the art to carry out the invention. (TRIPs, art. 29(1)). In addition to ensuring that the claimed invention in fact meets patent eligibility requirements, disclosure also permits third parties to access the invention for purposes of improving it or developing new inventions. In the case of plant-related innovations, national laws permit inventors to satisfy the disclosure requirement either by a written description of the invention and/or by a deposit of the protected material (i.e. seeds, germplasm or other biological material). TRIPs does not require national governments to adopt any particular form of disclosure, nor does it specify the timing, manner or conditions of third party access. As a result, national laws vary widely on this point.

Novelty, inventive step and industrial application : These three cumulative eligibility requirements are mandated by article 27(1) of the TRIPs Agreement and are found in all national patent laws. Even among states with strong patent protection laws, however, the interpretation and application of each of these steps varies considerably.

Novelty : The principal objective of the novelty requirement is to ensure that the claimed invention cannot be found in the "state of the art" or "prior art" already in existence. These terms refer to the body of knowledge which is available to the public before the date of filing of an application for patent protection. Patent examiners establish both the novelty and inventive step requirements by comparing the claimed invention to the state of the art or prior art. A critical issue for plant-related innovations concerns the form in which prior art exists and the difficulty of accessing traditional knowledge as prior art. Under United States patent law, for example, novelty can be negated if "the invention was patented or described in a printed publication in this or a foreign country or in public use or on sale in this country... " (35 U.S.C. Sec. 102(b)) The important point is description: foreign *uses* of claimed inventions are not considered as part of prior art.

The Patent Cooperation Treaty (PCT) adopts a similar approach, limiting prior art to "everything which has been made available to the public anywhere in the world by means of written disclosure." (Rule 33.1 of Regulations under the PCT) Thus, in the United States and in countries following the PCT, even where foreign plant-related innovations are known, used or disclosed in other than written form, it may be possible for the inventor to obtain patent protection.

Article 54(2) of the European Patent Convention adopts a different approach, defining the state of the art to encompass everything made available to the public by means of written or oral descriptions, by use or indeed in any other way, prior to the filing of the application. Nevertheless, the absence of systemic databases concerning traditional knowledge, including knowledge relating to plants, has led examiners in various countries to grant patents for inventions that were later revoked after they were discovered to encompass prior art.

Inventive step : The inventive step requirement looks to the state of the art to determine whether the claimed invention is not obvious to a person skilled in the art. Although the majority of international treaties and national laws defined inventive step in this way, the application of the standard to plant-related innovations varies and may be a significant hurdle to patent protection given "the enormous speed of technological progress" in the field.

Industrial application : The industrial application requirement is concerned with an invention's practical utility, and in particular whether "it can be made or used in any kind of industry, including agriculture." (E.g. European Patent Convention, art. 7) This eligibility requirement does not appear to pose a significant obstacle to the patenting of new plant varieties, given their ready use in the fields of plant breeding, horticulture and agriculture.

Exclusions from Patentability

As noted above, article 27.3(b) of TRIPs permits WTO Members to deny all patent protection to "plants" and "plant varieties." Nevertheless, a state that chooses to provide such protection may nevertheless choose to exclude from patentability a narrower category of inventions, "the prevention within their territory of the commercial exploitation of which is necessary to protect *ordre public* or morality, including to protect... plant life or health or to avoid serious prejudice to the environment." (Art. 27(2)) Some commentators have asserted that this language grants states the discretion to exclude broad categories of inventions that further the erosion of genetic diversity on the theory that issues of public policy and morality include a state's sovereign right to control its natural resources. However, because WTO dispute settlement panels have strictly interpreted clauses in the WTO Agreements that permit states to derogate from protected treaty rights, the more persuasive view is that exclusions on this basis must be fact-specific, narrowly drawn and no broader than required to achieve their purpose. (Moufang, 1998) Resolution of this debate in the context of plant-related innovations will not occur until the WTO dispute settlement bodies issue a ruling on the matter, an unlikely prospect so long as Members are free to exclude plants and plant varieties entirely from patentability.

Exclusive Rights

Once a patent is awarded, the owner of the patented product or process enjoys a broad panoply of exclusive rights. These include the right to prevent third parties from making the product, using the process or from using, offering for sale, selling or importing for those purposes the patented product or the product obtained by the patented process. (TRIPs, art. 28) States that choose to extend patent protection to plant-related innovations would thus be expected to protect products composed of plants and parts of plants and processes (including biological processes) for the production of plants. Although there is some dispute over whether traditional plant breeding methods (as opposed to creating varieties through genetic manipulation) can be protected by process patents, the practice of states that do extend patent protection to plants reveals that such methods have been protected.

Term of Protection

TRIPs article 33 imposes a minimum 20-year term of protection from the date the patent application is filed.

Exceptions and Limitations

As compared to limitations on plant breeders' rights permitted under the UPOV, the limitations on a patent owner's exclusive rights permitted under the TRIPs Agreement are far narrower. These limitations can be divided into exceptions to exclusive rights and compulsory licenses, which permit certain uses by third parties but require remuneration to the patent owner.

Exceptions : Article 30 of TRIPs permits WTO Members to adopt "limited exceptions" to exclusive patent rights provided that they do not "unreasonably conflict with a normal exploitation of the patent" and "do not unreasonably prejudice the legitimate interests of the patent owner, taking account of the legitimate interests of third parties." This three-part test is applicable to all patents protected by the TRIPs Agreement.

Canadian Patents Case : A WTO dispute settlement panel clarified the meaning of article 30 in Canada - Patent Protection of Pharmaceutical Products (Generic Medicines), WT/DS114/R (17 March 2000). The panel considered a complaint brought by the European Communities to two provisions of Canadian patent law. The first permitted uses of patented pharmaceuticals without the patentee's authorization for the purpose of obtaining approval of a generic product before the patent's term expired.

The second permitted production and stockpiling of the generic product for release immediately after the expiration of the patent. Both procedures were designed to permit the marketing of generic versions of pharmaceuticals promptly after the expiry of the patent. The panel decision confirmed that only the first exception is consistent with TRIPs. More importantly, the panel concluded that any exception that results in "a substantial curtailment" of the patent owner's exclusive rights is inconsistent with article 30. (WT/DS114/R, para. 7.36)

If TRIPs were amended to require patent protection for new plant varieties, the standard adopted by the Canadian Patents panel would preclude Members from enacting many of the exceptions that are permitted under plant variety protection laws, such as the breeders' exemption and the farmers' privilege. The same prohibition on these privileges already exists in several countries that currently protect plant varieties and other plant-related innovations in their national patent laws.

Research and experimentation exception : Most national patent laws permit third parties to engage in experimentation or research related to the patented invention. Some states, however, including the United States, narrowly construe this exception to preclude unauthorized commercialization of any products or processes that result from such research if they encompass the patentable invention. In such countries, commercial activities permitted under the breeders' exemption found in plant variety protection laws are prohibited where the variety is protected by a patent.

Thus, for example, a breeder would not be permitted to cross patented seeds to produce improved varieties. Even in patent systems in which the experimental use exception tolerates some commercialization (Cornish, 1998), such as those found in Europe, it is likely that the commercial acts allowed under a traditional breeders' exemption would "conflict with a normal exploitation of the patent" and would thus amount "a substantial curtailment" of his or her exclusive rights in violation of TRIPs article 30. Thus, if patent protection of plant varieties were to become mandatory under TRIPs, a breeders' exemption such as that found in article 22,V of Mexico's patent law would be unlikely to survive a challenge before a WTO dispute settlement panel.

Farmers' privilege : A similar fate is likely to befall the exception permitting farmers to save and reuse seeds on their own land without the patent owner's permission, another exception traditionally allowed under plant variety protection laws. The United States does not recognize a farmers' privilege under its utility patent laws. Although such an exception has been adopted in article 11 of the European Union's 1998 Biotechology Directive (conditioned upon the payment of equitable remuneration except by small farmers), at least one commentator has questioned whether that exception is compatible with TRIPs article 30, because such an exception would unreasonably prejudice the "legitimate interests" of the patent owner.

Compulsory licences : TRIPs contains a complex set of rules that regulate when states may compel patent owners to license their products and processes to governments or to private parties. Although the TRIPs Agreement does not specify the grounds which justify the creation of compulsory licences, because article 5A(2) of the Paris Convention is incorporated by reference into article 2(1) of TRIPs, it can be inferred that such licences may be granted only to prevent "abuses which might result from the exercise of the [patent owner's] exclusive rights."

Even where such abuses exist, TRIPs article 31 imposes further conditions upon the granting of compulsory licences, including individual consideration of each case, prior negotiations with the patent owner seeking a voluntary licence, limitations on the scope and duration of compulsory licences and requiring their termination when the circumstances leading to their creation are no longer in effect. Most importantly, patent owners must receive "adequate remuneration," taking into account the value of the rights licensed.

One area in which compulsory licences may affect plant breeders is that of dependent patents, which are defined as patents whose use requires the authorization of an earlier patent owner. Such patents are prevalent in plant breeding, where the creation of new varieties often occurs incrementally in the form of adaptations and improvements of existing varieties, as opposed to radically new innovations. Because incremental innovation often requires access to protected varieties, governments may seek to impose compulsory licences in favour of third party breeders who are unable to negotiate voluntary access to patented plant varieties. However, the compatibility of such licences with TRIPs is untested and governments seeking to impose them should ensure that they comply scrupulously with each of the many requirements set forth in article 31.

Because the patent compulsory licence provisions of TRIPs are far more detailed and narrow than the compulsory licences permitted under the 1978 and 1991 UPOV Acts, commentators are divided over their usefulness in the area of plant-related innovations. Some argue that such licences will be largely irrelevant to the field, whereas others claim that governments may adopt such licences "to ensure access to patented materials in order to attain specific agricultural objectives (e.g. availability of a given material for farmers) or food security...." Even if licences to achieve these objectives are permitted, however, the restrictions that TRIPs imposes significantly limit the ability of WTO Members to grant such licences in favour of third parties.

Sui Generis Protection for Plant Varieties under TRIPs

As noted above, TRIPs authorizes WTO Members to eschew patent protection for plants and plant varieties and adopt instead an "effective *sui generis* system" of protection. The following discussion identifies the core requirements of such a system, laying the groundwork for Part III which outlines the permissible options available to WTO Members seeking to grant IPRs to plant varieties while simultaneously seeking to achieve other objectives.

Identifying and Evaluating Competing Proposals for Sui Generis Systems

TRIPs does not define the term *sui generis* as it appears in article 27.3(b), nor does the treaty's drafting history shed any light on its meaning. As a general matter, the term *sui generis* is understood to mean "of its own kind" or "unique," a definition which does little to identify what sort of unique legal systems are permitted under the treaty. This lack of definitional clarity has allowed advocacy groups and commentators to advance different and often competing proposals for how effective *sui generis* protection should be structured, producing significant confusion for governments seeking to understand and implement their obligations under the TRIPs Agreement.

The discussion below identifies and evaluates the competing proposals. In brief, it concludes that if WTO Members choose to protect plant varieties using a *sui generis* right as opposed to a patent, they are required to create a distinct IPR applicable to plant varieties that complies with the core requirements and objectives of the TRIPs Agreement.

Sui generis systems that do not contain these core elements or conflict with them are not TRIPs-compatible and expose the WTO Members who adopt them to challenges under the WTO dispute settlement system. Nevertheless, the mandatory requirements of article 27.3(b) preserve significant leeway for national governments to work out the precise manner in which they will balance protection of IPRs against other international obligations and national objectives.

Sui Generis Systems as Defined by the UPOV Acts

Several commentators assert that a state adopting national legislation in compliance with either the 1978 UPOV Act or the 1991 UPOV Act has satisfied its obligations under article 27.3(b). Those advocating for compliance with the

1991 Act stress the more extensive IPR protections it provides to plant breeders, while commentators endorsing the 1978 Act as the appropriate *sui generis* standard stress that only the earlier Act was in force at the time TRIPs was being negotiated.

In fact, however, the protection required by the two UPOV Acts is neither necessary nor sufficient for an "effective *sui generis* system." It is not necessary because article 27.3(b) does not require plant variety protection laws to contain the same subject matter, eligibility requirements, exclusive rights, terms of protection or other detailed provisions of either of the two UPOV Acts. And it is not sufficient because TRIPs requires WTO Members to structure their national IPR laws in ways that the two UPOV Acts do not.

Note, however, that developing counties that are parties to a "TRIPs plus" bilateral or regional trade or investment treaty may be required, by virtue of those treaties, either to join UPOV or to implement the standards set forth in one of the UPOV Acts. A country that has undertaken such obligations does *not* possess the same flexibility in designing a *sui generis* system to protect plant varieties as a country which has not undertaken such obligations.

For example, as explained in paragraphs 2.2.2.4 and 2.2.3.4 above, both the 1991 Act and the 1978 Act adopt the principle of national treatment, the same principle required by TRIPs. But under the UPOV Acts such treatment need be extended only to the nationals and residents of *other UPOV member states* and to legal persons having their headquarters in such states. Any WTO Member that limited national treatment in this way would unquestionably violate article 3 of TRIPs, which requires extension of national treatment to all other WTO Members whether or not they are also UPOV members.

However, to comply with article 3, a state whose plant variety protection law contains a UPOV-specific national treatment rule can easily amend its law to expand the number of states to which national treatment applies. Other UPOV-sanctioned rules may require more extensive changes. For example, the 1978 Act allows a state to impose a reciprocity requirement in certain circumstances, limiting the right to apply for protection of a new variety to nationals or residents of those member states which also apply the Act to the genus or species to which that variety belongs. This reciprocal protection is incompatible with TRIPs' national treatment rule, and possibly its most favoured nation principle as well, both of which are inconsistent with a system of reciprocity.

That TRIPs' drafters did not intend either UPOV Act to be the exclusive model for *sui generis* protection of plant varieties is confirmed by their failure to refer to the Acts anywhere in the Agreement. By contrast, where the drafters intended Members to comply with standards found in preexisting international

IPR treaties, they stated so expressly. Nevertheless, as explained in sections 3.3.1 and 3.3.2 below, most provisions of the two UPOV Acts are fully consistent with an effective *sui generis* system, and most countries have in fact adopted plant variety protection laws that are consistent with one or both UPOV Acts.

Core Requirements of an Effective Sui Generis System

According to a 1997 report prepared by two leading commentators, there are four core elements that any national plant variety protection law must contain in order to qualify as an "effective *sui generis* system" within the meaning of TRIPs article 27.3(b):(1) the law must apply to all plant varieties in all species and botanical genera; (2) it must grant plant breeders an IPR, i.e. the exclusive right to control particular acts with respect to those protected varieties, or at a minimum, the right to remuneration when third parties engage in certain acts; (3) it must provide national treatment and MFN treatment to breeders from other WTO Members; and (4) it must contain procedures that enable breeders to enforce the rights granted to them under such a law. The paragraphs that follow discuss each of these four requirements.

Protection of all plant varieties : Although both UPOV Acts require protection of only a specific number of plant varieties (paras. 2.2.1.1 and 2.2.3.1 above), a careful reading of the TRIPs Agreement reveals that WTO Members are required to protect all varieties. article 27.3(b) states that "Members shall provide for the protection of *plant varieties*" (emphasis added) without any further qualification or limitation. Inasmuch as article 27.3(b) is included in a provision of the TRIPs Agreement which carves out numerous and specific exceptions to patent protection, if the drafters had intended to allow states to limit the number or type of plant varieties subject to protection, they would have said so expressly.

Indeed, the drafters were aware that both UPOV Acts permitted precisely such limitations, yet they chose neither to refer to those Acts nor to their limited scope of protection. As explained in paragraph 3.4.2.1 below, however, the obligation to protect all varieties does not require states to provide *the same level of protection* to each and every variety.

Protection of plant varieties with an IPR : If protection granted to plant varieties must take the form of an IPR, WTO Members would enjoy virtually unbounded discretion to choose the manner in which to protect plant varieties, for example through a system of taxation on seeds. The WTO's Appellate Body recently interpreted the text of the TRIPs Agreement, however, to require protection through an IPR. Nevertheless, because some commentators and NGOs have questioned whether plant variety protection is in fact a form of intellectual property included in the TRIPs Agreement, a brief discussion of this point is needed.

Article 1(2) of TRIPs defines “intellectual property” for purposes of the Agreement as “all categories of intellectual property that are the subject of Section 1 through 7 of Part II” of TRIPs. Protection for plant varieties appears in Section 5 relating to patents, and is thus a form of intellectual property protected by the Agreement. This interpretation is confirmed by the United States - Section 211 Omnibus Appropriations Act of 1998, AB-2001-7, WT/DS176/AB/R (2 January 2002). In that case, a WTO panel concluded that trade names were not a category of intellectual property protected by TRIPs because they were not expressly referred to in article 1(2) (an omission that also applies to plant varieties). The Appellate Body reversed this finding. As an example, it referred to Members’ option to protect plant varieties “by *sui generis* rights (such as breeder’s rights) instead of through patents,” and concluded that *sui generis* rights were in fact a form of intellectual property protected by the treaty. (Id., para. 335)

The significance of this conclusion is that IPRs for plant varieties must take the form of other IPRs protected by the TRIPs Agreement. Specifically, governments must either grant to the owners of protected varieties (1) the right to exclude all third parties from engaging in certain activities with respect to those varieties (an exclusive rights approach) or, (2) at a minimum, the right to receive equitable remuneration when a third party engages in such activities.

Nothing in TRIPs requires states to adopt the exclusive rights approach, nor does it indicate which exclusive rights must be granted to new varieties. Nevertheless, the vast majority of IPRs are protected by exclusive rights, with compulsory licences used only in limited areas such as new industries or distribution networks, abuses of monopoly powers or where high transaction costs preclude private licensing mechanisms from developing. In addition, a number of economic studies have suggested that granting exclusive rights to the owners of intellectual property products is a more efficient mechanism for creating incentives to innovate than a compulsory licence approach, in which political factors or government inertia are likely to produce skewed incentives. (Merges, 1996)

National and MFN treatment applicable to rights granted : At least one respected commentator has argued that national and MFN treatment are inapplicable to a *sui generis* system protecting plant varieties. The theory advanced is that these two forms of treatment are inapplicable to IPRs not covered by the TRIPs Agreement and that plant variety protection is an IPR that is not covered by TRIPs. However, the text of TRIPs article 3(1), as interpreted by WTO dispute settlement jurists, indicates that plant variety protection must adhere to such obligations.

Thus, in the United States - Section 211 Omnibus Appropriations Act of 1998, AB-2001-7, 2 January 2002, the WTO Appellate Body stressed the critical importance of the national and MFN treatment rules in multilateral trade agreements (paras. 242 and 297), and it concluded that the obligation to grant such treatment applies to all subjects of intellectual property protected by TRIPs, a designation that includes *sui generis* protection of plant varieties. As a result, in the area of plant variety protection, each Member must grant "no less favourable treatment" to the nationals of all other WTO Members than it grants to its own nationals and it must also grant to such foreign nationals "any advantage, favour, privilege or immunity" provided to any other country. (TRIPs, arts. 3.1 and 4) Applied to plant varieties, this would mean that any IPR that a state grants to its own plant breeders must be granted to breeders from all other WTO Members, and that any IPR granted to breeders from one WTO Member must be granted to breeders from all WTO Members.

Enforcement of rights by private parties : The only qualification that article 27.3(b) imposes upon *sui generis* systems for protecting plant varieties is that they be "effective." Although TRIPs does not define this term, it refers to it in Part III dealing with enforcement of other IPRs against acts of infringement by third parties. For this reason, a WTO dispute settlement panel is unlikely to find a *sui generis* system to be effective unless it provides a meaningful opportunity for private parties to enforce their rights in protected varieties.

Specific enforcement measures are not mandated by the Agreement, however. Had the drafters intended greater specificity, they could have easily made a reference in article 27.3(b) to the detailed enforcement rules applicable to other IPRs in Part III of TRIPs. Although a *sui generis* system must be "effective," this does not mean that a state must devote a particular level of resources to enforcement measures. In particular, article 41.5 of TRIPs makes clear that the treaty does not create "any obligation with respect to the distribution of resources as between enforcement of intellectual property rights and the enforcement of law in general." As a result, what qualifies as "effective" in a state with abundant resources and well-developed systems of law enforcement will not be required of states with more limited resources or weaker law enforcement systems.

Additional requirements of an "effective"* sui generis *system? : Beyond effective enforcement measures, it is uncertain whether a WTO dispute settlement panel would find that article 27.3(b) imposes additional requirements for a *sui generis* system to qualify as "effective." In general, international tribunals have concluded that rights granted in a treaty must be interpreted so as to make them effective rather than illusory. Thus, if a WTO Member were to grant very limited and weak rights to breeders together with broad and extensive exemptions in favour of farmers or other users of plant germplasm, its plant variety protection law might not survive WTO scrutiny. Such a possibility is likely to arise only in extreme cases in which a WTO Member has enacted a plant variety protection system that is functional in theory but non-functional in practice. States that implement the core TRIPs requirements in good faith - that is, states that grant breeders intellectual property rights and enforcement measures applicable to varieties in all species and botanical genera and that provide those same rights and measures to breeders from other WTO Members - are unlikely to have their laws challenged successfully.

Options Available to National Governments under Existing International IPR Agreements Protecting Plant Varieties and Plant Breeders' Rights

In order to identify the options available to national governments under international IPR agreements, this section begins by categorizing states according to the agreements to which they are parties. This classification scheme takes into account the fact that a state may be bound by more than one IPR-related treaty, a fact that may significantly limit its discretion. The section then identifies both the mandatory actions required of national governments in each category as well as the discretionary choices available to them. It also identifies trends in national laws which illustrate the manner in which states are complying with their particular set of international obligations.

Classification of States According to their International IPR Obligations

The discretion enjoyed by states to shape their plant variety protection laws to balance the protection of IPRs against other societal concerns is dependent upon the international agreement or agreements to which they are parties. The following table identifies that discretion at a broad level of generality. It does so by listing the mandatory components of plant variety protection that a state must adopt depending upon whether it is a member of (1) WTO only; (2) WTO and the 1991 UPOV Act; (3) WTO and the 1978 UPOV Act; and (4) either the 1991 UPOV Act or the 1978 UPOV Act only; or (5) no IPR agreement relating to the protection of plant varieties.

Note that some developing states, by virtue of having ratified a "TRIPs plus" bilateral or regional trade or investment treaty, may be required to adhere to some or all of the provisions contained in either the 1991 UPOV Act or the 1978 UPOV Act, *even if they are not members of UPOV itself.* Developing states that are parties to such "TRIPs plus" treaties should carefully consult the text of these agreements to determine the precise scope of their supplementary IPR obligations, including for purposes of applying the classification scheme set forth below. For example, a developing country that is a WTO Member but not a member of UPOV may nevertheless be required to adhere to all of the provisions of the 1991 UPOV Act by virtue of having ratified a "TRIPs plus" treaty. In such a case, the state would be obligated to implement the mandatory components of plant variety protection contained not only in TRIPs but also in the 1991 UPOV Act, and its discretion to balance the protection of IPRs against other competing policy objectives would be reduced accordingly.

Cumulative Treaty Obligations

Thus, for example, a state that is a party to a UPOV Act and also to the TRIPs Agreement must comply with its obligations under both treaties. Given that each treaty contains somewhat different but mostly cumulative and consistent rules concerning the protection of plant varieties, a state that is a party to both treaties will necessary have more extensive obligations to protect plant varieties and thus will enjoy concomitantly less discretion.

Conflicting Treaty Obligations

Where the provisions of two treaties are in direct conflict, the rule is far less settled. If two agreements relate to the same subject matter and the states parties to both agreements are the same, article 30 of the Vienna Convention on the Law of Treaties specifies that the agreement that is later in time is given effect. The obligations in the TRIPs Agreement are thus likely to prevail over any conflicting obligations in the 1978 UPOV Act, such as the ban on protecting varieties within the same genus or species with both a breeders' right and a patent. (1978 Act, art. 2(1)) Because the 1991 Act entered into force in 1998 and TRIPs entered into force in 1995 it may be argued that the 1991 UPOV Act is the later-in-time agreement. However, it is unlikely that any conflicts issue will arise between the two treaties, inasmuch as nothing in TRIPs requires conduct that the 1991 UPOV Act forbids. (Such a conflict might arise, however, in the unlikely event that TRIPs were interpreted to impose more restrictive exclusions to breeders' exclusive rights than the mandatory exclusions found in article 15 of the 1991 UPOV Act.) It can therefore be argued that the two treaty systems are fully compatible, with TRIPs merely augmenting the plant variety protection requirements of the UPOV Acts.

Identifying the Level of Discretion Available to States in each Treaty Classification

In general, the discretion a state enjoys in limiting the scope of intellectual property protection provided to plant varieties increases.

WTO and UPOV 1991 Act Members

States that are members of both TRIPs and the 1991 UPOV Act have the least discretion. This designation currently applies to 27 nations: Australia, Austria, Belarus, Bulgaria, Croatia, the Czech Republic, Denmark, Estonia,

Finland, Germany, Hungary, Israel, Japan, Kyrgyzstan, Latvia, Lithuania, the Netherlands, Moldova, Poland, Romania, Singapore, Slovenia, South Korea, Sweden, Tunisia, the United Kingdom and the United States of America.

Mandatory Requirements

These 27 states must extend protection to all plant varieties, comply with TRIPs' national MFN treatment obligations and adopt effective enforcement measures. The adoption of these rights (1) production or reproduction, (2) conditioning for the purpose of propagation, (3) offering for sale, (4) selling or marketing, (5) exporting, (6) importing and (7) stocking for any of these purposes - fully satisfies their obligation under TRIPs to protect varieties with an IPR. Finally, states that have agreed to these dual obligations must also comply with all of the other provisions of the 1991 Act, including its terms of protection and limitations on the breeders' exemption and farmers' privilege.

Options and Implementation Issues

A state whose plant variety protection laws are already in compliance with the 1991 Act needs to take only limited steps to comply with article 27.3(b) of TRIPs, including immediate protection of all varieties and extension of national treatment to all WTO Members. The most difficult hurdle is likely to enforce their rights. However, far more detailed enforcement measures must be adopted for the other forms of intellectual property protected by TRIPs. A state may therefore choose to extend to breeders the same domestic enforcement opportunities that it is required to provide to the owners of other IPRs.

National Implementation

In information submitted to the TRIPs Council in May 1999, the following states indicated that they had adopted plant variety protection laws modelled on the 1991 Act: Australia, Bulgaria, Members of the European Communities, Japan, Morocco, Poland, Romania, South Korea, South Africa, Switzerland and the United States of America. (Other states parties to both UPOV 1991 and TRIPs did not submit information to the Council.) Of the states adopting such laws, Morocco, South Africa and Switzerland had not ratified the 1991 Act.

Bilateral IPR Agreements and UPOV Membership

A separate issue concerns whether WTO Members will continue to join UPOV after the entry into force of the TRIPs Agreement. Some commentators predicted that UPOV membership would stagnate as states capitalized on the greater discretion provided to them under the *sui generis* option of article 27.3(b). Since TRIPs entered into force in 1995, however, 28 nations have become parties to the 1978 or 1991 UPOV Acts. (States Parties to the UPOV, Status on 30 June 2004) The continued expansion of UPOV membership can be traced in part to the recent proliferation of "TRIPs plus" treaties.

WTO and UPOV 1978 Act Members

This designation currently applies to 24 nations: Argentina, Bolivia, Brazil, Canada, Chile, China, Colombia, Ecuador, France, Ireland, Italy, Kenya, Mexico, New Zealand, Nicaragua, Norway, Panama, Paraguay, Portugal, Slovakia, South Africa, Switzerland, Trinidad and Tobago and Uruguay.

States in this category enjoy somewhat greater discretion as a result of the more limited protection of plant breeders' rights contained in this earlier UPOV Act.

Mandatory Requirements

States parties to both agreements must extend protection to all plant varieties, comply with TRIPs' national and MFN treatment rules and adopt effective enforcement measures.

They must also comply with all of the other 1978 Act requirements, including its eligibility requirements, terms of protection, exclusive rights and mandatory breeders' exemption. As compared to the 1991 Act, however, breeders' exclusive rights are more limited, terms of protection for varieties are shorter and exceptions and limitations are broader.

Options and Implementation Issues

States that become Members of the WTO after joining the 1978 UPOV Act and adopting laws to comply with that Act face a similar situation to states that are parties to both the WTO and the 1991 Act. To fully comply with TRIPs, these 1978 Act member states must modify their national laws to protect the four core requirements of article 27.3(b) and they must remove all provisions of their laws which impose a reciprocity requirement as a condition for protecting varieties of foreign breeders.

In addition, states in this category may choose to modify their laws to incorporate some or all of the standards found in the 1991 Act without actually becoming a member of that Act.

Their refusal to do so, however, does not violate article 27.3(b), inasmuch as the standards found in the 1978 Act satisfy their obligation to protect plant varieties with a *sui generis* IPR.

Trends in National Laws

In information submitted to the TRIPs Council in May 1999, Canada, the Czech Republic, Hungary, Norway and New Zealand (the only states submitting information) indicated that they had adopted plant variety protection laws that complied with the 1978 Act. At the time of these submissions, the Czech Republic and Hungary were members of the 1978 Act. They have since acceded to the 1991 Act.

1991 or 1978 UPOV Act Members Only

There are few states outside of the WTO that protect plant varieties solely by virtue of being UPOV members. And both are observer governments with the WTO. These nations face a somewhat different set of international legal obligations.

Mandatory Requirements

They must comply with all of the numerous requirements of the UPOV Act to which they are a party. However, they have no obligation to provide effective enforcement measures to breeders, their national treatment obligations are limited to those states that are members of the same UPOV Act and, in the case of those states bound by the 1978 Act, they need not protect all plant varieties and may impose certain reciprocity requirements on those varieties they do protect.

Options and Implementation Issues

Provided that their laws are in compliance with the relevant UPOV Act and unless they have ratified other international IPR agreements, these states need make no additional changes to their national plant variety protection laws.

Member of WTO Only

States that are parties to the TRIPs Agreement but are not members of either UPOV Act enjoy considerably greater discretion than those that are members of either UPOV Act alone. By virtue of their adherence to TRIPs, these states are required only to comply with the four core obligations of article 27.3(b). However, those countries that have ratified "TRIPs plus" treaties that require them to comply with one of the UPOV Acts).

Given the large number of states in this category (and the additional number of states likely to enter this category if their efforts to join the WTO succeed), a more detailed discussion of the numerous options available to them in balancing protection of plant breeders' rights against the competing policy objectives is provided in section 3.4 below.

Not a Member of WTO, UPOV or other IPR Agreements

Nations in this category have no obligation to protect plant varieties or breeders' rights in any form. Nevertheless, the mandatory requirements and discretionary choices facing other governments are likely to become matters of significant concern to these states as the world trading and intellectual property protection systems expand.

Attention to IPR issues will be of particular importance when these states implement the obligations of other international treaties to which they may be parties, such as the biodiversity goals of the CBD or the access requirements of the ITPGR. States in this category should consider whether it is sound policy to enact laws that are incompatible with TRIPs and the UPOV, since such an action will preclude their membership in these treaty systems unless they undertake a further revision of their national laws.

In addition, unilateral actions by or bilateral agreements with industrialized nations may pressure these states to comply with international IPR standards even if they remain outside of the WTO or UPOV.

For these reasons, states that are not yet Members of the WTO should give careful attention to the discretionary choices available to WTO Members analysed in the next section.

7

DNA Profiling

DNA profiling (also called DNA testing, DNA typing, or genetic fingerprinting) is a technique employed by forensic scientists to assist in the identification of individuals by their respective DNA profiles. DNA profiles are encrypted sets of numbers that reflect a person's DNA makeup, which can also be used as the person's identifier.

DNA profiling should not be confused with full genome sequencing. It is used in, for example, parental testing and rape investigation. Although 99.9% of human DNA sequences are the same in every person, enough of the DNA is different to distinguish one individual from another. DNA profiling uses repetitive ("repeat") sequences that are highly variable, called variable number tandem repeats (VNTR). VNTRs loci are very similar between closely related humans, but so variable that unrelated individuals are extremely unlikely to have the same VNTRs.

The DNA profiling technique was first reported in 1984 by Sir Alec Jeffreys at the University of Leicester in England, and is now the basis of several national DNA databases. Dr. Jeffreys's genetic fingerprinting was made commercially available in 1987, when a chemical company, ICI, started a blood-testing centre in England.

DNA Profiling Process

The process begins with a sample of an individual's DNA (typically called a "reference sample"). The most desirable method of collecting a reference sample is the use of a buccal swab, as this reduces the possibility of contamination. When this is not available (e.g. because a court order may be needed and not obtainable) other methods may need to be used to collect a sample of blood, saliva, semen, or other appropriate fluid or tissue from personal items (e.g. toothbrush, razor, etc.) or from stored samples (e.g. banked sperm or biopsy tissue). Samples obtained from blood relatives (biological relative) can provide an indication of an individual's profile, as could human remains which had been previously profiled. A reference sample is then analysed to create the individual's DNA profile using one of a number of techniques, discussed below. The DNA profile is then compared against another sample to determine whether there is a genetic match.

Restriction Fragment Length Polymorphism

In molecular biology, the term restriction fragment length polymorphism, or RFLP, (commonly pronounced "rif-lip") is a technique that exploits variations in homologous DNA sequences. It refers to a difference between samples of homologous DNA molecules that come from differing locations of restriction enzyme sites, and to a related laboratory technique by which these segments can be illustrated.

In RFLP analysis the DNA sample is broken into pieces (digested) by restriction enzymes and the resulting *restriction fragments* are separated according to their lengths by gel electrophoresis. Although now largely obsolete, RFLP analysis was the first DNA profiling technique inexpensive enough to see widespread application. In addition to genetic fingerprinting, RFLP was an important tool in genome mapping, localization of genes for genetic disorders, determination of risk for disease, and paternity testing.

Analysis Technique

The basic technique for detecting RFLPs involves fragmenting a sample of DNA by a restriction enzyme, which can recognize and cut DNA wherever a specific short sequence occurs, in a process known as a restriction digest. The resulting DNA fragments are then separated by length through a process known as agarose gel electrophoresis, and transferred to a membrane via the Southern blot procedure. Hybridization of the membrane to a labelled DNA probe then determines the length of the fragments which are complementary to the probe. An RFLP occurs when the length of a detected fragment varies between individuals. Each fragment length is considered an allele, and can be used in genetic analysis.

RFLP analysis may be subdivided into single- (SLP) and multi-locus probe (MLP) paradigms. Usually, the SLP method is preferred over MLP because it is more sensitive, easier to interpret and capable of analyzing mixed-DNA samples. Moreover data can be generated even when the DNA is degraded (e.g. when it is found in bone remains.)

Examples

There are two common mechanisms by which the size of a particular restriction fragment can vary. In the first schematic, a small segment of the genome is being detected by a DNA probe (thicker line). In allele "A", the genome is cleaved by a restriction enzyme at three nearby sites (triangles), but only the rightmost fragment will be detected by the probe. In allele "a", restriction site 2 has been lost by a mutation, so the probe now detects the larger fused fragment running from sites 1 to 3. The second diagram shows how this fragment size variation would look on a Southern blot, and how each allele (two per individual) might be inherited in members of a family.

In the third schematic, the probe and restriction enzyme are chosen to detect a region of the genome that includes a variable VNTR segment (boxes). In allele "c" there are five repeats in the VNTR, and the probe detects a longer fragment between the two restriction sites. In allele "d" there are only two repeats in the VNTR, so the probe detects a shorter fragment between the same two restriction sites. Other genetic processes, such as insertions, deletions, translocations, and inversions, can also lead to RFLPs.

Applications

Analysis of RFLP variation in genomes was a vital tool in genome mapping and genetic disease analysis. If researchers were trying to initially determine the chromosomal location of a particular disease gene, they would analyse the DNA of members of a family afflicted by the disease, and look for RFLP alleles that show a similar pattern of inheritance as that of the disease. Once a disease gene was localized, RFLP analysis of other families could reveal who was at risk for the disease, or who was likely to be carriers of the mutant genes.

RFLP analysis was also the basis for early methods of Genetic fingerprinting, useful in the identification of samples retrieved from crime scenes, in the determination of paternity, and in the characterization of genetic diversity or breeding patterns in animal populations.

Alternatives

The technique for RFLP analysis is, however, slow and cumbersome. It requires a large amount of sample DNA, and the combined process of probe labeling, DNA fragmentation, electrophoresis, blotting, hybridization, washing, and autoradiography could take up to a month to complete. A limited version of the RFLP method that used oligonucleotide probes was reported in 1985. Fortunately, the results of the Human Genome Project have largely replaced the need for RFLP mapping, and the identification of many single-nucleotide polymorphisms (SNPs) in that project (as well as the direct identification of many disease genes and mutations) has replaced the need for RFLP disease linkage analysis. The analysis of VNTR alleles continues, but is now usually performed by polymerase chain reaction (PCR) methods. For example, the standard protocols for DNA fingerprinting involve PCR analysis of panels of more than a dozen VNTRs.

RFLP is still a technique used in marker assisted selection. Terminal restriction fragment length polymorphism (TRFLP or sometimes T-RFLP) is a molecular biology technique initially developed for characterizing bacterial communities in mixed-species samples. The technique has also been applied to other groups including soil fungi.

TRFLP works by PCR amplification of DNA using primer pairs that have been labelled with fluorescent tags. The PCR products are then digested using RFLP enzymes and the resulting patterns visualized using a DNA sequencer. The results are analysed either by simply counting and comparing bands or peaks in the TRFLP profile, or by matching bands from one or more TRFLP runs to a database of known species. The technique is similar in some aspects to DGGE or TGGE.

The sequence changes directly involved with an RFLP can also be analysed more quickly by PCR. Amplification can be directed across the altered restriction site, and the products digested with the restriction enzyme. This method has been called Cleaved Amplified Polymorphic Sequence (CAPS). Alternatively, the amplified segment can by analysed by Allele specific oligonucleotide (ASO) probes, a process that can often be done by a simple Dot blot.

Short Tandem Repeat

A short tandem repeat (STR) in DNA occurs when a pattern of two or more nucleotides are repeated and the repeated sequences are directly adjacent to each other. The pattern can range in length from 2 to 50 base pairs (bp) (for example $(CATG)_n$ in a genomic region) and is typically in the non-coding intron region. A short tandem repeat polymorphism (STRP) occurs when homologous STR loci differ in the number of repeats between individuals.

By identifying repeats of a specific sequence at specific locations in the genome, it is possible to create a genetic profile of an individual. There are currently over 10,000 published STR sequences in the human genome. STR analysis has become the prevalent analysis method for determining genetic profiles in forensic cases.

Forensic STR Analysis

STR analysis is a relatively new technology in the field of forensics, having come into popularity in the mid-to-late 1990s. It is used for the genetic fingerprinting of individuals. The STRs in use today for forensic analysis are all tetra- or penta-nucleotide repeats (4 or 5 repeated nucleotides), as these give a high degree of error-free data while being robust enough to survive degradation in non-ideal conditions. Shorter repeat sequences tend to suffer from artifacts such as PCR stutter and preferential amplification, as well as the fact that several genetic diseases are associated with tri-nucleotide repeats such as Huntington's disease. Longer repeat sequences will suffer more highly from environmental degradation and do not amplify by PCR as well as shorter sequences. The analysis is performed by extracting nuclear DNA from the cells of a forensic sample of interest, then amplifying specific polymorphic regions of the extracted DNA by means of the polymerase chain reaction.

Once these sequences have been amplified, they are resolved either through gel electrophoresis or capillary electrophoresis, which will allow the analyst to determine how many repeats of the STR sequence in question there are. If the DNA was resolved by gel electrophoresis, the DNA can be visualized either by silver staining (not very high resolution, safe, inexpensive), or an intercalating dye such as ethidium bromide (fairly sensitive, moderate health risks, inexpensive), or as most modern forensics labs use, fluorescent dyes (highly sensitive, safe, expensive).

Instruments built to resolve STR fragments by capillary electrophoresis also use fluorescent dyes to great effect. It is also used to follow up bone marrow transplant patients. In the United States, 13 core STR loci have been decided upon to be the basis by which an individual genetic profile can be generated. These profiles are stored on a local, state and national level in DNA databanks such as CODIS. The British data base for STR loci identification is the UK National DNA Database (NDNAD). The British system uses 10 loci, rather than the American 13 loci. Y-STRs (STRs on the Y chromosome) are often used in genealogical DNA testing.

Amplified Fragment Length Polymorphism

Amplified Fragment Length Polymorphism PCR (or AFLP-PCR or just AFLP) is a PCR-based tool used in genetics research, DNA fingerprinting, and in the practice of genetic engineering. Developed in the early 1990s by Keygene, AFLP uses restriction enzymes to digest genomic DNA, followed by ligation of adaptors to the sticky ends of the restriction fragments. A subset of the restriction fragments is then selected to be amplified.

This selection is achieved by using primers complementary to the adaptor sequence, the restriction site sequence and a few nucleotides inside the restriction site fragments (as described in detail below). The amplified fragments are visualized on denaturing polyacrylamide gels either through autoradiography or fluorescence methodologies.

AFLP-PCR is a highly sensitive method for detecting polymorphisms in DNA. The technique was originally described by Vos and Zabeau in 1993. In detail, the procedure of this technique is divided into three steps:

1. Digestion of total cellular DNA with one or more restriction enzymes and ligation of restriction half-site specific adaptors to all restriction fragments.
2. Selective amplification of some of these fragments with two PCR primers that have corresponding adaptor and restriction site specific sequences.

3. Electrophoretic separation of amplicons on a gel matrix, followed by visualisation of the band pattern.

A variation on AFLP is cDNA-AFLP, which is used to quantify differences in gene expression levels. Another variation on AFLP is TE Display, used to detect transposable element mobility.

Applications

The AFLP technology has the capability to detect various polymorphisms in different genomic regions simultaneously. It is also highly sensitive and reproducible. As a result, AFLP has become widely used for the identification of genetic variation in strains or closely related species of plants, fungi, animals, and bacteria. The AFLP technology has been used in criminal and paternity tests, also to determine slight differences within populations, and in linkage studies to generate maps for quantitative trait locus (QTL) analysis.

There are many advantages to AFLP when compared to other marker technologies including randomly amplified polymorphic DNA (RAPD), restriction fragment length polymorphism (RFLP), and microsatellites. AFLP not only has higher reproducibility, resolution, and sensitivity at the whole genome level compared to other techniques, but it also has the capability to amplify between 50 and 100 fragments at one time. In addition, no prior sequence information is needed for amplification (Meudt & Clarke 2007). As a result, AFLP has become extremely beneficial in the study of taxa including bacteria, fungi, and plants, where much is still unknown about the genomic makeup of various organisms.

The AFLP technology is covered by patents and patent applications of Keygene N.V. AFLP is a registered trademark of Keygene N.V.

DNA Family Relationship Analysis

Using PCR technology, DNA analysis is widely applied to determine genetic family relationships such as paternity, maternity, siblingship and other kinships. During conception, the father's sperm cell and the mother's egg cell, each containing half the amount of DNA found in other body cells, meet and fuse to form a fertilized egg, called a zygote. The zygote contains a complete set of DNA molecules, a unique combination of DNA from both parents. This zygote divides and multiplies into an embryo and later, a full human being.

At each stage of development, all the cells forming the body contain the same DNA—half from the father and half from the mother. This fact allows the relationship testing to use all types of all samples including loose cells from the cheeks collected using buccal swabs, blood or other types of samples.

While a lot of DNA contains information for a certain function, there is some called junk DNA, which is currently used for human identification. At some special locations (called loci) in the junk DNA, predictable inheritance patterns were found to be useful in determining biological relationships. These locations contain specific DNA markers that DNA scientists use to identify individuals. In a routine DNA paternity test, the markers used are Short Tandem Repeats (STRs), short pieces of DNA that occur in highly differential repeat patterns among individuals.

Each person's DNA contains two copies of these markers—one copy inherited from the father and one from the mother. Within a population, the markers at each person's DNA location could differ in length and sometimes sequence, depending on the markers inherited from the parents.

The combination of marker sizes found in each person makes up his/her unique genetic profile. When determining the relationship between two individuals, their genetic profiles are compared to see if they share the same inheritance patterns at a statistically conclusive rate.

For example, the following sample report from this commercial DNA paternity testing laboratory Universal Genetics signifies how relatedness between parents and child is identified on those special markers:

DNA Marker	*Mother*	*Child*	*Alleged father*
D21S11	28, 30	28, 31	29, 31
D7S820	9, 10	10, 11	11, 12
TH01	14, 15	14, 16	15, 16
D13S317	7, 8	7, 9	8, 9
D19S433	14, 16.2	14, 15	15, 17

The partial results indicate that the child and the alleged father's DNA match among these five markers. The complete test results show this correlation on 16 markers between the child and the tested man to draw a conclusion of whether or not the man is the biological father.

Scientifically, each marker is assigned with a Paternity Index (PI), which is a statistical measure of how powerfully a match at a particular marker indicates paternity.

The PI of each marker is multiplied with each other to generate the Combined Paternity Index (CPI), which indicates the overall probability of an individual being the biological father of the tested child relative to any random man from the entire population of the same race.

The CPI is then converted into a Probability of Paternity showing the degree of relatedness between the alleged father and child.

The DNA test report in other family relationship tests, such as grandparentage and siblingship tests, is similar to a paternity test report. Instead of the Combined Paternity Index, a different value, such as a Siblingship Index, is reported.

The report shows the genetic profiles of each tested person. If there are markers shared among the tested individuals, the probability of biological relationship is calculated to determine how likely the tested individuals share the same markers due to a blood relationship.

Y-chromosome Analysis

Recent innovations have included the creation of primers targeting polymorphic regions on the Y-chromosome (Y-STR), which allows resolution of a mixed DNA sample from a male and female and/or cases in which a differential extraction is not possible. Y-chromosomes are paternally inherited, so Y-STR analysis can help in the identification of paternally related males. Y-STR analysis was performed in the Sally Hemings controversy to determine if Thomas Jefferson had sired a son with one of his slaves.

Mitochondrial DNA

Replication

mtDNA is replicated by the DNA polymerase gamma complex which is composed of a 140 kDa catalytic DNA polymerase encoded by the *POLG* gene and a 55 kDa accessory subunit encoded by the *POLG2* gene. During embryogenesis, replication of mtDNA is strictly down-regulated from the fertilized oocyte through the preimplantation embryo. At the blastocyst stage, the onset of mtDNA replication is specific to the cells of the trophectoderm. In contrast, the cells of the inner cell mass restrict mtDNA replication until they receive the signals to differentiate to specific cell types.

Origin

Nuclear and mitochondrial DNA are thought to be of separate evolutionary origin, with the mtDNA being derived from the circular genomes of the bacteria that were engulfed by the early ancestors of today's eukaryotic cells. This theory is called the endosymbiotic theory.

Each mitochondrion is estimated to contain 2-10 mtDNA copies. In the cells of extant organisms, the vast majority of the proteins present in the mitochondria (numbering approximately 1500 different types in mammals) are coded for by nuclear DNA, but the genes for some of them, if not most, are thought to have originally been of bacterial origin, having since been transferred to the eukaryotic nucleus during evolution.

Mitochondrial Inheritance

In most multicellular organisms, mtDNA is inherited from the mother (maternally inherited). Mechanisms for this include simple dilution (an egg contains 100,000 to 1,000,000 mtDNA molecules, whereas a sperm contains only 100 to 1000), degradation of sperm mtDNA in the fertilized egg, and, at least in a few organisms, failure of sperm mtDNA to enter the egg. Whatever the mechanism, this single parent (uniparental) pattern of mtDNA inheritance is found in most animals, most plants and in fungi as well.

Female Inheritance

In sexual reproduction, mitochondria are normally inherited exclusively from the mother. The mitochondria in mammalian sperm are usually destroyed by the egg cell after fertilization. Also, most mitochondria are present at the base of the sperm's tail, which is used for propelling the sperm cells. Sometimes the tail is lost during fertilization. In 1999 it was reported that paternal sperm mitochondria (containing mtDNA) are marked with ubiquitin to select them for later destruction inside the embryo. Some *in vitro* fertilization techniques, particularly injecting a sperm into an oocyte, may interfere with this.

The fact that mitochondrial DNA is maternally inherited enables researchers to trace maternal lineage far back in time. (Y chromosomal DNA, paternally inherited, is used in an analogous way to trace the agnate lineage.) This is accomplished in humans by sequencing one or more of the hypervariable control regions (HVR1 or HVR2) of the mitochondrial DNA, as with a genealogical DNA test. HVR1 consists of about 440 base pairs.

These 440 base pairs are then compared to the control regions of other individuals (either specific people or subjects in a database) to determine maternal lineage. Most often, the comparison is made to the revised Cambridge Reference Sequence. Vilà *et al.* have published studies tracing the matrilineal descent of domestic dogs to wolves. The concept of the Mitochondrial Eve is based on the same type of analysis, attempting to discover the origin of humanity by tracking the lineage back in time.

Because mtDNA is not highly conserved and has a rapid mutation rate, it is useful for studying the evolutionary relationships - phylogeny - of organisms. Biologists can determine and then compare mtDNA sequences among different species and use the comparisons to build an evolutionary tree for the species examined. Because mtDNA is transmitted from mother to child (both male and female), it can be a useful tool in genealogical research into a person's maternal line.

Paternal mtDNA Transmission

In genetics, paternal mtDNA transmission and paternal mtDNA inheritance refer to the incidence of mitochondrial DNA (mtDNA) being passed from a father to his offspring. Paternal mtDNA inheritance is observed in a small proportion of species; in general, mtDNA is passed unchanged from a mother to her offspring, making it an example of non-Mendelian inheritance. In contrast, mtDNA transmission from both parents occurs regularly in certain bivalves.

In Animals

Paternal mtDNA inheritance in animals varies. For example, in Mytilidae mussels, paternal mtDNA "is transmitted through the sperm and establishes itself only in the male gonad." In testing 172 sheep, "The Mitochondrial DNA from three lambs in two half-sib families were found to show paternal inheritance."

In Humans

In human mitochondrial genetics, there is debate over whether or not paternal mtDNA transmission is possible. Many studies hold that paternal mtDNA is never transmitted to offspring. This belief is central to mtDNA genealogical DNA testing and to the theory of mitochondrial Eve. The fact that mitochondrial DNA is maternally inherited enables researchers to trace maternal lineage far back in time. (Y chromosomal DNA, paternally inherited, is used in an analogous way to trace the agnate lineage.)

In sexual reproduction, paternal mitochondria found in the sperm are actively decomposed, thus preventing "paternal leakage". Mitochondria in mammalian sperm are usually destroyed by the egg cell after fertilization. In 1999 it was reported that paternal sperm mitochondria (containing mtDNA) are marked with ubiquitin to select them for later destruction inside the embryo. Some in vitro fertilization (IVF) techniques, particularly intracytoplasmic sperm injection (ICSI) of a sperm into an oocyte, may interfere with this. It is now understood that the tail of the sperm, which contains additional mtDNA, may also enter the egg. This had led to increased controversy about the fate of paternal mtDNA.

Some sources state that so little paternal mtDNA is transmitted as to be negligible ("At most, one presumes it must be less than 1 in 1000, since there are 100 000 mitochondria in the human egg and only 100 in the sperm (Satoh and Kuroiwa, 1991).") or that paternal mtDNA is so rarely transmitted as to be negligible ("Nevertheless, studies have established that paternal mtDNA is so rarely transmitted to offspring that mtDNA analyses remain valid..."). One study stated that about 1–2% of a person's mitochondria can be inherited from the father.

The controversy about human paternal leakage was summed up in the 1996 study *Misconceptions about mitochondria and mammalian fertilization: Implications for theories on human evolution*. The following quotation comes from the abstract to that peer-reviewed study printed in the *Proceedings of the National Academy of Sciences*: The mixing of maternal and paternal mtDNA was thought to have been found in humans and chimpanzees in 1999. However, there has been only a single documented case of human paternal mitochondrial DNA transmission, and it was linked to infertility.

According to the 2005 study *More evidence for non-maternal inheritance of mitochondrial DNA?*, heteroplasmy is a "newly discovered form of inheritance for mtDNA. Heteroplasmy introduces slight statistical uncertainty in normal inheritance patterns." Heteroplasmy may result from a mutation during development which is propagated to only a subset of the adult cells, or may occur when two slightly different mitochondrial sequences are inherited from the mother as a result of several hundred mitochondria being present in the ovum.

In Protozoa

Some organisms, such as *Cryptosporidium*, have mitochondria with no DNA whatsoever.

In Plants

In plants, it has also been reported that mitochondria can occasionally be inherited from the father, e.g. in bananas. Some Conifers also show paternal inheritance of mitochondria, such as the coast redwood, *Sequoia sempervirens*.

Structure

In humans (and probably in metazoans in general), 100-10,000 separate copies of mtDNA are usually present per cell (egg and sperm cells are exceptions). In mammals, each double-stranded circular mtDNA molecule consists of 15,000-17,000 base pairs. The two strands of mtDNA are differentiated by their nucleotide content with the guanine rich strand referred to as the heavy strand, and the cytosine rich strand referred to as the light strand. The heavy strand encodes 28 genes, and the light strand encodes 9 genes for a total of 37 genes. Of the 37 genes, 13 are for proteins (polypeptides), 22 are for transfer RNA (tRNA) and two are for the small and large subunits of ribosomal RNA (rRNA). This pattern is also seen among most metazoans, although in some cases one or more of the 37 genes is absent and the mtDNA size range is greater. Even greater variation in mtDNA gene content and size exists among fungi and plants, although there appears to be a core subset of genes that are present in all eukaryotes (except for the few that have no mitochondria at all). Some plant species have enormous mtDNAs (as many as 2,500,000 base pairs per mtDNA molecule) but, surprisingly, even those huge mtDNAs contain the same number and kinds of genes as related plants with much smaller mtDNAs.

National DNA Database

A national DNA database is a government database of DNA profiles which can be used by law enforcement agencies to identify suspects of crimes. The first government datatabase (NDNAD) was set up by the United Kingdom in April 1995. The second one was set up in New Zealand. France set up the FNAEG in 1998. In the USA, the FBI has organized the CODIS database. Originally intended for sex offenders, they have since been extended to include almost any criminal offender. In England and Wales, anyone arrested on suspicion of a recordable offence must submit a DNA sample, the profile of which is then stored on the DNA database as a permanent record. In Scotland, the law requires the DNA profiles of most people who are acquitted be removed from the database.

In Sweden, only the DNA profiles of criminals who have spent more than two years in prison are stored. In Norway and Germany, court orders are required, and are only available, respectively, for serious offenders and for those convicted of certain offences and who are likely to reoffend. All 50 states in the USA store DNA profiles of violent offenders, and a few also store profiles of suspects. Portugal has plans to introduce a DNA database of its entire population

In Denmark the Danish Newborn Screening Biobank at Statens Serum Institut keeps a blood sample from people born after 1981. The purpose is to test for Phenylketonuria and other diseases. But it is also used for DNA tests to identify diseased and suspected criminals. Parents can request that the blood sample of their newborn is destroyed after the result of test is known.

Scope of Databases

The United States maintains the largest DNA database in the world, with the Combined DNA Index System (CODIS) holding over 5 million records as of 2007. The United Kingdom maintains the National DNA Database (NDNAD), which is of similar size. The size of this database, and its rate of growth, is giving concern to civil liberties and political groups in the UK, where police have wide-ranging powers to take samples and retain them even in the event of acquittal. When a match is made from a national DNA database to link a crime scene to an offender who has provided a DNA sample to a database that link is often referred to as a *cold hit*. A cold hit is of value in referring the police agency to a specific suspect but is of less evidential value than a DNA match made from outside the DNA database..

As of May 2007, 177,870 forensic profiles and 4,582,516 offender profiles have been accumulated, making it the largest DNA database in the world, surpassing the United Kingdom National DNA Database, which consisted of an estimated 3,976,090 profiles as of June 2007. As of the same date, CODIS has produced over 49,400 matches to requests, assisting in more than 50,343 investigations.

The growing public approval of DNA databases has seen the creation and expansion of many states' own DNA databases. California currently maintains the third largest DNA database in the world (naturally, as CODIS contains all states' database information). Political measures such as California Proposition 69 (2004), which increased the scope of the DNA database, have already met with a significant increase in numbers of investigations aided.

In order to decrease the number of irrelevant matches at NDIS, the Convicted Offender Index requires all 13 CODIS STRs to be present for a profile upload. Forensic profiles only require 10 of the STRs to be present for an upload.

DNA Databases and Medicine

The database became the common meeting ground for computer scientists and molecular biologists. This is because the goal of certain projects like the genome project was to construct maps, which were built from information contributed to databases. They enabled an entirely new way of information analysis. Many types of modern molecular biological analysis require computational analysis in parallel. However, the information gained from mapping and sequencing genetic information would very likely have ethical implications for individuals, families and society in general. There is a concern about genetic information being used in ways that affect chances of employment or chance of getting life insurance. Furthermore, secondary applications of personal, genetic information mean that citizens do not know what their genetic information will be used for.

DNA Collection and Human Rights

In a judgement in December 2008, the European Court of Human Rights ruled that two British men should not have had their DNA and fingerprints retained by police saying that retention "could not be regarded as necessary in a democratic society".

The DNA fingerprinting pioneer Professor Sir Alec Jeffreys condemned UK government plans to keep the genetic details of hundreds of thousands of innocent people in England and Wales for up to 12 years. Jeffreys said he was "disappointed" with the proposals, which came after a European court ruled that the current policy breaches people's right to privacy. Jefferys said "It seems to be as about as minimal a response to the European court of human rights judgment as one could conceive. There is a presumption not of innocence but of future guilt here ... which I find very disturbing indeed".

Considerations when Evaluating DNA Evidence

In the early days of the use of genetic fingerprinting as criminal evidence, juries were often swayed by spurious statistical arguments by defence lawyers along these lines: given a match that had a 1 in 5 million probability of occurring by chance, the lawyer would argue that this meant that in a country of say 60 million people there were 12 people who would also match the profile. This was then translated to a 1 in 12 chance of the suspect being the guilty one. This argument is not sound unless the suspect was drawn at random from the population of the country. In fact, a jury should consider how likely it is that an individual matching the genetic profile would also have been a suspect in the case for other reasons. Another spurious statistical argument is based on the false assumption that a 1 in 5 million probability of a match automatically translates into a 1 in 5 million probability of innocence and is known as the prosecutor's fallacy.

When using RFLP, the theoretical risk of a coincidental match is 1 in 100 billion (100,000,000,000), although the practical risk is actually 1 in 1000 because monozygotic twins are 0.2% of the human population. Moreover, the rate of laboratory error is almost certainly higher than this, and often actual laboratory procedures do not reflect the theory under which the coincidence probabilities were computed. For example, the coincidence probabilities may be calculated based on the probabilities that markers in two samples have bands in *precisely* the same location, but a laboratory worker may conclude that similar—but not precisely identical—band patterns result from identical genetic samples with some imperfection in the agarose gel. However, in this case, the laboratory worker increases the coincidence risk by expanding the criteria for declaring a match. Recent studies have quoted relatively high error rates which may be cause for concern. In the early days of genetic fingerprinting, the necessary population data to accurately compute a match probability was sometimes unavailable. Between 1992 and 1996, arbitrary low ceilings were controversially put on match probabilities used in RFLP analysis rather than the higher theoretically computed ones. Today, RFLP has become widely disused due to the advent of more discriminating, sensitive and easier technologies.

STRs do not suffer from such subjectivity and provide similar power of discrimination (1 in 10^13 for unrelated individuals if using a full SGM+ profile) It should be noted that figures of this magnitude are not considered to be statistically supportable by scientists in the UK, for unrelated individuals with full matching DNA profiles a match probability of 1 in a billion is considered statistically supportable (Since 1998 the DNA profiling system supported by The National DNA Database in the UK is the SGM+ DNA profiling system which includes 10 STR regions and a sex indicating test. However, with any DNA technique, the cautious juror should not convict on genetic fingerprint evidence alone if other factors raise doubt. Contamination with other evidence (secondary transfer) is a key source of incorrect DNA profiles and raising doubts as to whether a sample has been adulterated is a favorite defence technique. More rarely, chimerism is one such instance where the lack of a genetic match may unfairly exclude a suspect.

Evidence of Genetic Relationship

It's also possible to use DNA profiling as evidence of genetic relationship, but testing that shows no relationship isn't absolutely certain. While almost all individuals have a single and distinct set of genes, rare individuals, known as "chimeras", have at least two different sets of genes. There have been several cases of DNA profiling that falsely "proved" that a mother was unrelated to her children.

Fake DNA Evidence

The value of DNA evidence has to be seen in light of recent cases where criminals planted fake DNA samples at crime scenes. In one case, a criminal even planted fake DNA evidence in his own body: Dr. John Schneeberger raped one of his sedated patients in 1992 and left semen on her underwear. Police drew what they believed to be Schneeberger's blood and compared its DNA against the crime scene semen DNA on three occasions, never showing a match. It turned out that he had surgically inserted a Penrose drain into his arm and filled it with foreign blood and anticoagulants.

In a study conducted by the life science company Nucleix and published in the journal Forensic Science International, scientists found that an In vitro synthesized sample of DNA matching any desired genetic profile can be constructed using standard molecular biology techniques without obtaining any actual tissue from that person.

In the case of the Phantom of Heilbronn, police detectives found DNA traces from the same woman on various crime scenes in Austria, Germany and France - among them murders, burglaries and robberies. Only after the DNA of the "woman" matched the DNA sampled from the burned body of a *male* asylum seeker in France, detectives began to have serious doubts about the DNA evidence. In that case, DNA traces were already present on the cotton swabs used to collect the samples at the crime scene, and the swabs had all been produced at the same factory in Austria. The company's product specification said that the swabs were guaranteed to be sterile, but not DNA-free.

Familial Searching

Familial searching is the use of family members' DNA to identify a closely related suspect in jurisdictions where large DNA databases exist, but no exact match has been found. The first successful use of the practice was in a UK case where a man was convicted of manslaughter when he threw a brick stained with his own blood into a moving car. Police could not get an exact match to the UK's DNA database because the man had no criminal convictions, but police implicated him using a close relative's DNA. The technique was used to catch a Los Angeles serial killer known as the "Grim Sleeper" in 2010.

Surreptitious DNA Collecting

Police forces may collect DNA samples without the suspects' knowledge, and use it as evidence. Legality of this mode of proceeding has been questioned in Australia.

In the United States, it has been accepted, courts often claiming that there was no expectation of privacy, citing California v. Greenwood (1985), during which the Supreme Court held that the Fourth Amendment does not prohibit the warrantless search and seizure of garbage left for collection outside the curtilage of a home. Critics of this practice underline the fact that this analogy ignores that "most people have no idea that they risk surrendering their genetic identity to the police by, for instance, failing to destroy a used coffee cup. Moreover, even if they do realize it, there is no way to avoid abandoning one's DNA in public."

In the UK, the Human Tissue Act 2004 prohibited private individuals from covertly collecting biological samples (hair, fingernails, etc.) for DNA analysis, but excluded medical and criminal investigations from the offense.

England and Wales

Evidence from an expert who has compared DNA samples must be accompanied by evidence as to the sources of the samples and the procedures for obtaining the DNA profiles. The judge must ensure that the jury must understand the significance of DNA matches and mismatches in the profiles.

The judge must also ensure that the jury does not confuse the 'match probability' (the probability that a person that is chosen at random has a matching DNA profile to the sample from the scene) with the 'likelihood ratio' (the probability that a person with matching DNA committed the crime). In *R v. Doheny* Phillips LJ gave this example of a summing up, which should be carefully tailored to the particular facts in each case:

Members of the Jury, if you accept the scientific evidence called by the Crown, this indicates that there are probably only four or five white males in the United Kingdom from whom that semen stain could have come. The Defendant is one of them. If that is the position, the decision you have to reach, on all the evidence, is whether you are sure that it was the Defendant who left that stain or whether it is possible that it was one of that other small group of men who share the same DNA characteristics.

Juries should weigh up conflicting and corroborative evidence, using their own common sense and not by using mathematical formulae, such as Bayes' theorem, so as to avoid "confusion, misunderstanding and misjudgment".

Presentation and Evaluation of Evidence of Partial or Incomplete DNA Profiles

In *R v Bates*, Moore-Bick LJ said:

> *"We can see no reason why partial profile DNA evidence should not be admissible provided that the jury are made aware of its inherent limitations and are given a sufficient explanation to enable them to evaluate it. There may be cases where the match*

probability in relation to all the samples tested is so great that the judge would consider its probative value to be minimal and decide to exclude the evidence in the exercise of his discretion, but this gives rise to no new question of principle and can be left for decision on a case by case basis. However, the fact that there exists in the case of all partial profile evidence the possibility that a "missing" allele might exculpate the accused altogether does not provide sufficient grounds for rejecting such evidence. In many there is a possibility (at least in theory) that evidence exists which would assist the accused and perhaps even exculpate him altogether, but that does not provide grounds for excluding relevant evidence that is available and otherwise admissible, though it does make it important to ensure that the jury are given sufficient information to enable them to evaluate that evidence properly".

DNA Testing in the US

There are state laws on DNA profiling in all 50 states of the United States. Detailed information on database laws in each state can be found at the National Conference of State Legislatures website.

Development of Artificial DNA

In August 2009, scientists in Israel stunned the forensic sciences and raised serious questions concerning the use of DNA by law enforcement as the ultimate method of identification. In a paper published in the journal *Forensic Science International: Genetics*, the Israeli researchers demonstrated that it is possible to manufacture DNA in a laboratory, and thus falsify DNA evidence. The scientists fabricated saliva and blood samples, which originally contained DNA from a person other than the ostensible donor of the blood and saliva.

The researchers also showed that, using a DNA database, it is possible to take information from a profile and manufacture DNA to match it, and that this can be done without access to any actual DNA from the person whose DNA they are duplicating. The synthetic DNA oligos required for the procedure are common in molecular laboratories.

Dr. Daniel Frumkin, lead author on the paper, was quoted in *The New York Times* as saying, "You can just engineer a crime scene... any biology undergraduate could perform this." Dr. Frumkin perfected a test that can forensically differentiate real DNA samples from fake ones. His test detects epigenetic modifications, in particular, DNA methylation.

Seventy percent of the DNA in any human genome is methylated, meaning it contains methyl group modifications within a CpG dinucleotide context. Methylation at the promoter region is associated with gene silencing. The synthetic DNA lacks this epigenetic modification, which allows the test to distinguish manufactured DNA from original, genuine, DNA. It is unknown how many, if any, police departments currently use the test, which appears to be a serious issue in light of Frumkin's claim that the DNA manufacturing procedure is within the grasp of any undergraduate biology student. No police lab has publicly announced that it is using the new test to verify DNA results, while FSI Genetics says that any forensic laboratory doing DNA identification should adopt this test to authenticate its results as "real" DNA.

Cases

- In the 1950s, Anna Anderson claimed that she was Grand Duchess Anastasia Nikolaevna of Russia; in the 1980s after her death, samples of her tissue that had been stored at a Charlottesville, Virginia hospital following a medical procedure were tested using DNA fingerprinting and showed that she bore no relation to the Romanovs.
- In 1986, Richard Buckland was exonerated despite having admitted to the rape and murder of a teenager near Leicester, the city where DNA profiling was first discovered. This was the first use of DNA finger printing in a criminal investigation.
- In 1987, in the same case as Buckland, British baker Colin Pitchfork was the first criminal caught and convicted using DNA fingerprinting.
- In 1987, genetic fingerprinting was used in criminal court for the first time in the trial of a man accused of unlawful intercourse with a mentally handicapped 14-year-old female who gave birth to his baby.
- In 1987, Florida rapist Tommy Lee Andrews was the first person in the United States to be convicted as a result of DNA evidence, for raping a woman during a burglary; he was convicted on 6 November 1987 and sentenced to 22 years in prison.
- In 1988, Timothy Spencer was the first man in Virginia to be sentenced to death through DNA testing, for several rape and murder charges; he was dubbed "The South Side Strangler" because he killed victims on the southside of Richmond, Virginia. He was later charged with rape and 1st degree murder and was sentenced to death. He was executed on April 27, 1994. David Vasquez, initially convicted of one of Spencer's crimes, became the first man in America exonerated based on DNA evidence.
- In 1989, Chicago man Gary Dotson was the first person whose conviction was overturned using DNA evidence.
- In 1991, Allan Legere was the first Canadian to be convicted as a result of DNA evidence, for four murders he had committed while an escaped prisoner in 1989. During his trial, his defence argued that the relatively shallow gene pool of the region could lead to false positives.

- In 1992, DNA evidence was used to prove that Nazi doctor Josef Mengele was buried in Brazil under the name Wolfgang Gerhard.
- In 1993, Kirk Bloodsworth was the first person to have been convicted of murder and sentenced to death, whose conviction was overturned using DNA evidence.
- The 1993 rape and murder of Mia Zapata, lead singer for the Seattle punk band The Gits was unsolved 9 years after the murder. A database search in 2001 failed, but the killer's DNA was collected when he was arrested in Florida for burglary and domestic abuse in 2002.
- The science was made famous in the United States in 1994 when prosecutors heavily relied on — and through expert witnesses exhaustively presented and explained — DNA evidence allegedly linking O.J. Simpson to a double murder. The case also brought to light the laboratory difficulties and handling procedure mishaps which can cause such evidence to be significantly doubted.
- In 1994, Royal Canadian Mounted Police (RCMP) detectives successfully tested hairs from a cat known as Snowball, and used the test to link a man to the murder of his wife, thus marking for the first time in forensic history the use of non-human DNA to identify a criminal.
- In 1998, Dr. Richard J. Schmidt was convicted of attempted second-degree murder when it was shown that there was a link between the viral DNA of the human immunodeficiency virus (HIV) he had been accused of injecting in his girlfriend and viral DNA from one of his patients with full-blown AIDS. This was the first time viral DNA fingerprinting had been used as evidence in a criminal trial.
- In 1999, Raymond Easton a disabled man from Swindon, England was arrested and detained for 7 hours in connection with a burglary due to an inaccurrate DNA match. His DNA had been retained on file after an unrelated domestic incident some time previously.
- In May 2000 Gordon Graham murdered Paul Gault at his home in Lisburn, Northern Ireland. Graham was convicted of the murder when his DNA was found on a sports bag left in the house as part of an elaborate ploy to suggest the murder occurred after a burglary had gone wrong. Graham was having an affair with the victims wife at the time of the murder. It was the first time Low Copy Number DNA was used in Northern Ireland.
- In 2001, Wayne Butler was convicted for the murder of Celia Douty. It was the first murder in Australia to be solved using DNA profiling.
- In 2002, DNA testing was used to exonerate Douglas Echols, a man who was wrongfully convicted in a 1986 rape case. Echols was the 114th person to be exonerated through post-conviction DNA testing.

- In August 2002, Annalisa Vincenzi was shot dead in Tuscany. Some time later, Bartender Peter Hamkin, 23, was arrested in Merseyside in March 2003 on an extradition warrant heard at Bow Street Magistrates' Court in London to establish whether he should be taken to Italy to face a murder charge. DNA "proved" he shot her, but he was cleared on other evidence.
- In 2003, Welshman Jeffrey Gafoor was convicted of the 1988 murder of Lynette White, when crime scene evidence collected 12 years earlier was re-examined using STR techniques, resulting in a match with his nephew. This may be the first known example of the DNA of an innocent yet related individual being used to identify the actual criminal, via "familial searching".
- In June 2003, because of new DNA evidence, Dennis Halstead, John Kogut and John Restivo won a re-trial on their murder conviction. The three men had already served eighteen years of their thirty-plus-year sentences.
- The trial of Robert Pickton is notable in that DNA evidence is being used primarily to identify the *victims*, and in many cases to prove their existence.
- In March 2003, Josiah Sutton was released from prison after serving four years of a twelve-year sentence for a sexual assault charge. Questionable DNA samples taken from Sutton were retested in the wake of the Houston Police Department's crime lab scandal of mishandling DNA evidence.
- In 2004, DNA testing shed new light into the mysterious 1912 disappearance of Bobby Dunbar, a four-year-old boy who vanished during a fishing trip. He was allegedly found alive eight months later in the custody of William Cantwell Walters, but another woman claimed that the boy was her son, Bruce Anderson, whom she had entrusted in Walters' custody. The courts disbelieved her claim and convicted Walters for the kidnapping. The boy was raised and known as Bobby Dunbar throughout the rest of his life. However, DNA tests on Dunbar's son and nephew revealed the two were not related, thus establishing that the boy found in 1912 was not Bobby Dunbar, whose real fate remains unknown.
- In 2005, Gary Leiterman was convicted of the 1969 murder of Jane Mixer, a law student at the University of Michigan, after DNA found on Mixer's pantyhose was matched to Leiterman. DNA in a drop of blood on Mixer's hand was matched to John Ruelas, who was only four years old in 1969 and was never successfully connected to the case in any other way. Leiterman's defence unsuccessfully argued that the

unexplained match of the blood spot to Ruelas pointed to cross-contamination and raised doubts about the reliability of the lab's identification of Leiterman.

- In December 2005, Evan Simmons was proven innocent of a 1981 attack on an Atlanta woman after serving twenty-four years in prison. Mr Clark is the 164th person in the United States and the fifth in Georgia to be freed using post-conviction DNA testing.
- In March 2009, Sean Hodgson who spent 27 years in jail, convicted of killing Teresa De Simone, 22, in her car in Southampton 30 years ago was quashed by senior judges. Tests prove DNA from the scene was not his. British police have now reopened the case.

Bioinformatics

Bioinformatics is the application of statistics and computer science to the field of molecular biology. The term *bioinformatics* was coined by Paulien Hogeweg and Ben Hesper in 1978 for the study of informatic processes in biotic systems. Its primary use since at least the late 1980s has been in genomics and genetics, particularly in those areas of genomics involving large-scale DNA sequencing.

Bioinformatics now entails the creation and advancement of databases, algorithms, computational and statistical techniques and theory to solve formal and practical problems arising from the management and analysis of biological data. Over the past few decades rapid developments in genomic and other molecular research technologies and developments in information technologies have combined to produce a tremendous amount of information related to molecular biology. It is the name given to these mathematical and computing approaches used to glean understanding of biological processes.

Common activities in bioinformatics include mapping and analyzing DNA and protein sequences, aligning different DNA and protein sequences to compare them and creating and viewing 3-D models of protein structures.

The primary goal of bioinformatics is to increase the understanding of biological processes. What sets it apart from other approaches, however, is its focus on developing and applying computationally intensive techniques (e.g., pattern recognition, data mining, machine learning algorithms, and visualization) to achieve this goal. Major research efforts in the field include sequence alignment, gene finding, genome assembly, drug design, drug discovery, protein structure alignment, protein structure prediction, prediction of gene expression and protein-protein interactions, genome-wide association studies and the modelling of evolution. Bioinformatics was applied in the creation and maintenance of a database to store biological information at the beginning of the "genomic revolution", such as nucleotide and amino acid sequences. Development of this type of database involved not only design issues but the development of complex interfaces whereby researchers could both access existing data as well as submit new or revised data.

In order to study how normal cellular activities are altered in different disease states, the biological data must be combined to form a comprehensive picture of these activities. Therefore, the field of bioinformatics has evolved such that the most pressing task now involves the analysis and interpretation of various types of data, including nucleotide and amino acid sequences, protein domains, and protein structures. The actual process of analyzing and interpreting data is referred to as computational biology. Important sub-disciplines within bioinformatics and computational biology include:

- the development and implementation of tools that enable efficient access to, and use and management of, various types of information.
- the development of new algorithms (mathematical formulas) and statistics with which to assess relationships among members of large data sets, such as methods to locate a gene within a sequence, predict protein structure and/or function, and cluster protein sequences into families of related sequences.

There are two fundamental ways of modelling a Biological system (e.g. living cell) both coming under Bioinformatic approaches.

- Static
  - Sequences - Proteins, Nucleic acids and Peptides
  - Structures - Proteins, Nucleic acids, Ligands (including metabolites and drugs) and Peptides
  - Interaction data among the above entities including microarray data and Networks of proteins, metabolites
- Dynamic
  - Systems Biology comes under this category including reaction fluxes and variable concentrations of metabolites
  - Multi-Agent Based modelling approaches capturing cellular events such as signalling, transcription and reaction dynamics.

A broad sub-category under bioinformatics is structural bioinformatics.

Major Research Areas

Sequence Analysis

Since the Phage Φ-X174 was sequenced in 1977, the DNA sequences of thousands of organisms have been decoded and stored in databases. This sequence information is analysed to determine genes that encode polypeptides (proteins), RNA genes, regulatory sequences, structural motifs, and repetitive sequences. A comparison of genes within a species or between different species can show similarities between protein functions, or relations between species (the use of molecular systematics to construct phylogenetic trees). With the growing amount of data, it long ago became impractical to analyse DNA sequences manually. Today, computer programs such as BLAST are used daily to search sequences from more than 260 000 organisms, containing over 190 billion nucleotides. These programs can compensate for mutations (exchanged, deleted or inserted bases) in the DNA sequence, in order to identify sequences that are related, but not identical. A variant of this sequence alignment is used in the sequencing process itself.

The so-called shotgun sequencing technique (which was used, for example, by The Institute for Genomic Research to sequence the first bacterial genome, *Haemophilus influenzae*) does not produce entire chromosomes, but instead generates the sequences of many thousands of small DNA fragments (ranging from 35 to 900 nucleotides long, depending on the sequencing technology). The ends of these fragments overlap and, when aligned properly by a genome assembly program, can be used to reconstruct the complete genome. Shotgun sequencing yields sequence data quickly, but the task of assembling the fragments can be quite complicated for larger genomes. For a genome as large as the human genome, it may take many days of CPU time on large-memory, multiprocessor computers to assemble the fragments, and the resulting assembly will usually contain numerous gaps that have to be filled in later. Shotgun sequencing is the method of choice for virtually all genomes sequenced today, and genome assembly algorithms are a critical area of bioinformatics research.

Another aspect of bioinformatics in sequence analysis is annotation, which involves computational gene finding to search for protein-coding genes, RNA genes, and other functional sequences within a genome. Not all of the nucleotides within a genome are part of genes. Within the genome of higher organisms, large parts of the DNA do not serve any obvious purpose. This so-called junk DNA may, however, contain unrecognized functional elements. Bioinformatics helps to bridge the gap between genome and proteome projects — for example, in the use of DNA sequences for protein identification.

Gene Prediction

Gene finding typically refers to the area of computational biology that is concerned with algorithmically identifying stretches of sequence, usually genomic DNA, that are biologically functional. This especially includes protein-coding genes, but may also include other functional elements such as RNA genes and regulatory regions. Gene finding is one of the first and most important steps in understanding the genome of a species once it has been sequenced. In its earliest days, "gene finding" was based on painstaking experimentation on living cells and organisms. Statistical analysis of the rates of homologous recombination of several different genes could determine their order on a certain chromosome, and information from many such experiments could be combined to create a genetic map specifying the rough location of known genes relative to each other. Today, with comprehensive genome sequence and powerful computational resources at the disposal of the research community, gene finding has been redefined as a largely computational problem. Determining that a sequence *is functional* should be distinguished from determining *the function* of the gene or its product. The latter still demands *in vivo* experimentation through gene knockout and other assays, although frontiers of bioinformatics research are making it increasingly possible to predict the function of a gene based on its sequence alone.

Extrinsic Approaches

In extrinsic (or evidence-based) gene finding systems, the target genome is searched for sequences that are similar to extrinsic evidence in the form of the known sequence of a messenger RNA (mRNA) or protein product. Given an mRNA sequence, it is trivial to derive a unique genomic DNA sequence from which it had to have been transcribed. Given a protein sequence, a family of possible coding DNA sequences can be derived by reverse translation of the genetic code. Once candidate DNA sequences have been determined, it is a relatively straightforward algorithmic problem to efficiently search a target genome for matches, complete or partial, and exact or inexact. BLAST is a widely used system designed for this purpose.

A high degree of similarity to a known messenger RNA or protein product is strong evidence that a region of a target genome is a protein-coding gene. However, to apply this approach systemically requires extensive sequencing of mRNA and protein products. Not only is this expensive, but in complex organisms, only a subset of all genes in the organism's genome are expressed at any given time, meaning that extrinsic evidence for many genes is not readily accessible in any single cell culture. Thus, in order to collect extrinsic evidence for most or all of the genes in a complex organism, many hundreds or thousands of different cell types must be studied, which itself presents further difficulties. For example, some human genes may be expressed only during development as an embryo or fetus, which might be difficult to study for ethical reasons.

Despite these difficulties, extensive transcript and protein sequence databases have been generated for human as well as other important model organisms in biology, such as mice and yeast. For example, the RefSeq database contains transcript and protein sequence from many different species, and the Ensembl system comprehensively maps this evidence to human and several other genomes. It is, however, likely that these databases are both incomplete and contain small but significant amounts of erroneous data..

Ab Initio Approaches

Because of the inherent expense and difficulty in obtaining extrinsic evidence for many genes, it is also necessary to resort to *Ab initio* gene finding, in which genomic DNA sequence alone is systematically searched for certain tell-tale signs of protein-coding genes. These signs can be broadly categorized as either *signals*, specific sequences that indicate the presence of a gene nearby, or *content*, statistical properties of protein-coding sequence itself. *Ab initio* gene finding might be more accurately characterized as gene *prediction*, since extrinsic evidence is generally required to conclusively establish that a putative gene is functional.

In the genomes of prokaryotes, genes have specific and relatively well-understood promoter sequences (signals), such as the Pribnow box and transcription factor binding sites, which are easy to systematically identify. Also, the sequence coding for a protein occurs as one contiguous open reading frame (ORF), which is typically many hundred or thousands of base pairs long. The statistics of stop codons are such that even finding an open reading frame of this length is a fairly informative sign. (Since 3 of the 64 possible codons in the genetic code are stop codons, one would expect a stop codon approximately every 20-25 codons, or 60-75 base pairs, in a random sequence.) Furthermore, protein-coding DNA has certain periodicities and other statistical properties that are easy to detect in sequence of this length. These characteristics make prokaryotic gene finding relatively straightforward, and well-designed systems are able to achieve high levels of accuracy.

Ab initio gene finding in eukaryotes, especially complex organisms like humans, is considerably more challenging for several reasons. First, the promoter and other regulatory signals in these genomes are more complex and less well-understood than in prokaryotes, making them more difficult to reliably recognize. Two classic examples of signals identified by eukaryotic gene finders are CpG islands and binding sites for a poly(A) tail.

Second, splicing mechanisms employed by eukaryotic cells mean that a particular protein-coding sequence in the genome is divided into several parts (exons), separated by non-coding sequences (introns). (Splice sites are themselves another signal that eukaryotic gene finders are often designed to identify.) A typical protein-coding gene in humans might be divided into a dozen exons, each less than two hundred base pairs in length, and some as short as twenty to thirty. It is therefore much more difficult to detect periodicities and other known content properties of protein-coding DNA in eukaryotes.

Advanced gene finders for both prokaryotic and eukaryotic genomes typically use complex probabilistic models, such as hidden Markov models, in order to combine information from a variety of different signal and content measurements. The GLIMMER system is a widely used and highly accurate gene finder for prokaryotes. GeneMark is another popular approach. Eukaryotic *ab initio* gene finders, by comparison, have achieved only limited success; notable examples are the GENSCAN and geneid programs. The SNAP gene finder is HMM-based like Genscan and attempts to be more adaptable to different organisms, addressing problems related to using a gene finder on a genome sequence that it was not trained against.A few recent approaches like mSplicer, CONTRAST, or mGene also use machine learning techniques like support vector machines for successful gene prediction. They build a discriminative model using hidden Markov support vector machines or conditional random fields to learn an accurate gene prediction scoring function.

Other Signals

Among the derived signals used for prediction are statistics resulting from the sub-sequence statistics like k-mer statistics, Fourier transform of a pseudo-number-coded DNA, Z-curve parameters and certain run features. It has been suggested that signals other than those directly detectable in sequences may improve gene prediction. For example, the role of secondary structure in the identification of regulatory motifs has been reported. In addition, it has been suggested that RNA secondary structure prediction helps splice site prediction.

Comparative Genomics Approaches

As the entire genomes of many different species are sequenced, a promising direction in current research on gene finding is a comparative genomics approach. This is based on the principle that the forces of natural selection cause genes and other functional elements to undergo mutation at a slower rate than the rest of the genome, since mutations in functional elements are more likely to negatively impact the organism than mutations elsewhere. Genes can thus be detected by comparing the genomes of related species to detect this evolutionary pressure for conservation. This approach was first applied to the mouse and human genomes, using programs such as SLAM, SGP and Twinscan/N-SCAN.

Comparative gene finding can also be used to project high quality annotations from one genome to another. Notable examples include Projector, GeneWise and GeneMapper. Such techniques now play a central role in the annotation of all genomes.

Computational Evolutionary Biology

Evolutionary biology is the study of the origin and descent of species, as well as their change over time. Informatics has assisted evolutionary biologists in several key ways; it has enabled researchers to:

- trace the evolution of a large number of organisms by measuring changes in their DNA, rather than through physical taxonomy or physiological observations alone,

- more recently, compare entire genomes, which permits the study of more complex evolutionary events, such as gene duplication, horizontal gene transfer, and the prediction of factors important in bacterial speciation,
- build complex computational models of populations to predict the outcome of the system over time
- track and share information on an increasingly large number of species and organisms.

Future work endeavours to reconstruct the now more complex tree of life. The area of research within computer science that uses genetic algorithms is sometimes confused with computational evolutionary biology, but the two areas are not necessarily related.

Analysis of Gene Expression

The expression of many genes can be determined by measuring mRNA levels with multiple techniques including microarrays, expressed cDNA sequence tag (EST) sequencing, serial analysis of gene expression (SAGE) tag sequencing, massively parallel signature sequencing (MPSS), or various applications of multiplexed in-situ hybridization. All of these techniques are extremely noise-prone and/or subject to bias in the biological measurement, and a major research area in computational biology involves developing statistical tools to separate signal from noise in high-throughput gene expression studies. Such studies are often used to determine the genes implicated in a disorder: one might compare microarray data from cancerous epithelial cells to data from non-cancerous cells to determine the transcripts that are up-regulated and down-regulated in a particular population of cancer cells.

Analysis of Regulation

Regulation is the complex orchestration of events starting with an extracellular signal such as a hormone and leading to an increase or decrease in the activity of one or more proteins. Bioinformatics techniques have been applied to explore various steps in this process. For example, promoter analysis involves the identification and study of sequence motifs in the DNA surrounding the coding region of a gene. These motifs influence the extent to which that region is transcribed into mRNA. Expression data can be used to infer gene regulation: one might compare microarray data from a wide variety of states of an organism to form hypotheses about the genes involved in each state. In a single-cell organism, one might compare stages of the cell cycle, along with various stress conditions (heat shock, starvation, etc.). One can then apply clustering algorithms to that expression data to determine which genes are co-expressed. For example, the upstream regions (promoters) of co-expressed genes can be searched for over-represented regulatory elements.

Analysis of Protein Expression

Protein microarrays and high throughput (HT) mass spectrometry (MS) can provide a snapshot of the proteins present in a biological sample. Bioinformatics is very much involved in making sense of protein microarray and HT MS data; the former approach faces similar problems as with microarrays targeted at mRNA, the latter involves the problem of matching large amounts of mass data against predicted masses from protein sequence databases, and the complicated statistical analysis of samples where multiple, but incomplete peptides from each protein are detected.

Analysis of Mutations in Cancer

In cancer, the genomes of affected cells are rearranged in complex or even unpredictable ways. Massive sequencing efforts are used to identify previously unknown point mutations in a variety of genes in cancer. Bioinformaticians continue to produce specialized automated systems to manage the sheer volume of sequence data produced, and they create new algorithms and software to compare the sequencing results to the growing collection of human genome sequences and germline polymorphisms. New physical detection technologies are employed, such as oligonucleotide microarrays to identify chromosomal gains and losses (called comparative genomic hybridization), and single-nucleotide polymorphism arrays to detect known *point mutations*. These detection methods simultaneously measure several hundred thousand sites throughout the genome, and when used in high-throughput to measure thousands of samples, generate terabytes of data per experiment. Again the massive amounts and new types of data generate new opportunities for bioinformaticians. The data is often found to contain considerable variability, or noise, and thus Hidden Markov model and change-point analysis methods are being developed to infer real copy number changes.

Another type of data that requires novel informatics development is the analysis of lesions found to be recurrent among many tumours.

Comparative Genomics

Comparative genomics is the study of the relationship of genome structure and function across different biological species or strains. Comparative genomics is an attempt to take advantage of the information provided by the signatures of selection to understand the function and evolutionary processes that act on genomes. While it is still a young field, it holds great promise to yield insights into many aspects of the evolution of modern species. The sheer amount of information contained in modern genomes (3.2 gigabases in the case of humans) necessitates that the methods of comparative genomics are automated. Gene finding is an important application of comparative genomics, as is discovery of new, non-coding functional elements of the genome.

Comparative genomics exploits both similarities and differences in the proteins, RNA, and regulatory regions of different organisms to infer how selection has acted upon these elements. Those elements that are responsible for similarities between different species should be conserved through time (stabilizing selection), while those elements responsible for differences among species should be divergent (positive selection). Finally, those elements that are unimportant to the evolutionary success of the organism will be unconserved (selection is neutral).

One of the important goals of the field is the identification of the mechanisms of eukaryotic genome evolution. It is however often complicated by the multiplicity of events that have taken place throughout the history of individual lineages, leaving only distorted and superimposed traces in the genome of each living organism. For this reason comparative genomics studies of small model organisms (for example yeast) are of great importance to advance our understanding of general mechanisms of evolution.

Having come a long way from its initial use of finding functional proteins, comparative genomics is now concentrating on finding regulatory regions and siRNA molecules. Recently, it has been discovered that distantly related species often share long conserved stretches of DNA that do not appear to code for any protein. One such ultra-conserved region, that was stable from chicken to chimp has undergone a sudden burst of change in the human lineage, and is found to be active in the developing brain of the human embryo.

Computational approaches to genome comparison have recently become a common research topic in computer science. A public collection of case studies and demonstrations is growing, ranging from whole genome comparisons to gene expression analysis. This has increased the introduction of different ideas, including concepts from systems and control, information theory, strings analysis and data mining. It is anticipated that computational approaches will become and remain a standard topic for research and teaching, while multiple courses will begin training students to be fluent in both topics.

Modelling Biological Systems

Modelling biological systems is a significant task of systems biology and mathematical biology. Computational systems biology aims to develop and use efficient algorithms, data structures, visualization and communication tools to orchestrate the integration of large quantities of biological data with the goal of computer modelling.

It involves the use of computer simulations of biological systems, like cellular subsystems (such as the networks of metabolites and enzymes which comprise metabolism, signal transduction pathways and gene regulatory networks) to both analyse and visualize the complex connections of these cellular processes.

It is also directly associated with bioinformatics and computational biology. Artificial life or virtual evolution attempts to understand evolutionary processes via the computer simulation of simple (artificial) life forms.

Overview

It is understood that an unexpected emergent property of a complex system is a result of the interplay of the cause-and-effect among simpler, integrated parts. Biological systems manifest many important examples of emergent properties in the complex interplay of components. Traditional study of biological systems requires reductive methods in which quantities of data are gathered by category, such as concentration over time in response to a certain stimulus. Computers are critical to analysis and modelling of these data. The goal is to create accurate real-time models of a system's response to environmental and internal stimuli, such as a model of a cancer cell in order to find weaknesses in its signalling pathways, or modelling of ion channel mutations to see effects on cardiomyocytes and in turn, the function of a beating heart.

A monograph on this topic summarizes an extensive amount of published research in this area up to 1987, including subsections in the following areas: computer modelling in biology and medicine, arterial system models, neuron models, biochemical and oscillation networks, quantum automata, quantum computers in molecular biology and genetics, cancer modelling, neural nets, genetic networks, abstract relational biology, metabolic-replication systems, category theory applications in biology and medicine, automata theory, cellular automata, tessallation models and complete self-reproduction, chaotic systems in organisms, relational biology and organismic theories. This published report also includes 390 references to peer-reviewed articles by a large number of authors.

Standards

By far the most widely accepted standard format for storing and exchanging models in the field is the Systems Biology Markup Language (SBML) The SBML.org website includes a guide to many important software packages used in computational systems biology. Other markup languages with different emphases include BioPAX and CellML.

Particular Tasks

Cellular Model

Creating a cellular model has been a particularly challenging task of systems biology and mathematical biology. It involves developing efficient algorithms, data structures, visualization and communication tools to orchestrate the integration of large quantities of biological data with the goal of computer modelling.

It is also directly associated with bioinformatics, computational biology and Artificial life. It involves the use of computer simulations of the many cellular subsystems such as the networks of metabolites and enzymes which comprise metabolism, signal transduction pathways and gene regulatory networks to both analyse and visualize the complex connections of these cellular processes.

The complex network of biochemical reaction/transport processes and their spatial organization make the development of a predictive model of a living cell a grand challenge for the 21st century.

Overview

The eukaryotic cell cycle is very complex and is one of the most studied topics, since its misregulation leads to cancers. It is possibly a good example of a mathematical model as it deals with simple calculus but gives valid results. Two research groups have produced several models of the cell cycle simulating several organisms. They have recently produced a generic eukaryotic cell cycle model which can represent a particular eukaryote depending on the values of the parameters, demonstrating that the idiosyncrasies of the individual cell cycles are due to different protein concentrations and affinities, while the underlying mechanisms are conserved (Csikasz-Nagy et al., 2006).

By means of a system of ordinary differential equations these models show the change in time (dynamical system) of the protein inside a single typical cell; this type of model is called a deterministic process (whereas a model describing a statistical distribution of protein concentrations in a population of cells is called a stochastic process).

To obtain these equations an iterative series of steps must be done: first the several models and observations are combined to form a consensus diagram and the appropriate kinetic laws are chosen to write the differential equations, such as rate kinetics for stoichiometric reactions, Michaelis-Menten kinetics for enzyme substrate reactions and Goldbeter–Koshland kinetics for ultrasensitive transcription factors, afterwards the parameters of the equations (rate constants, enzyme efficiency coefficients and Michealis constants) must be fitted to match observations; when they cannot be fitted the kinetic equation is revised and when that is not possible the wiring diagram is modified. The parameters are fitted and validated using observations of both wild type and mutants, such as protein half-life and cell size.

In order to fit the parameters the differential equations need to be studied. This can be done either by simulation or by analysis. In a simulation, given a starting vector (list of the values of the variables), the progression of the system is calculated by solving the equations at each time-frame in small increments. In analysis, the proprieties of the equations are used to investigate the Behaviour of the system depending of the values of the parameters and variables. A system of differential equations can be represented as a vector field, where each vector described the change (in concentration of two or more protein) determining where and how fast the trajectory (simulation) is heading. Vector fields can have several special points: a stable point, called a sink, that attracts in all directions (forcing the concentrations to be at a certain value), an unstable point, either a source or a saddle point which repels (forcing the concentrations to change away from a certain value), and a limit cycle, a closed trajectory towards which several trajectories spiral towards (making the concentrations oscillate).

A better representation which can handle the large number of variables and parameters is called a bifurcation diagram (bifurcation theory): the presence of these special steady-state points at certain values of a parameter (e.g. mass) is represented by a point and once the parameter passes a certain value, a qualitative change occurs, called a bifurcation, in which the nature of the space changes, with profound consequences for the protein concentrations: the cell cycle has phases (partially corresponding to G1 and G2) in which mass, via a stable point, controls cyclin levels, and phases (S and M phases) in which the concentrations change independently, but once the phase has changed at a bifurcation event (cell cycle checkpoint), the system cannot go back to the previous levels since at the current mass the vector field is profoundly different and the mass cannot be reversed back through the bifurcation event, making a checkpoint irreversible. In particular the S and M checkpoints are regulated by means of special bifurcations called a Hopf bifurcation and an infinite period bifurcation.

8

Genetic Material: Properties and Replication

DNA

Deoxyribonucleic acid (DNA) is a nucleic acid that contains the genetic instructions used in the development and functioning of all known living organisms with the exception of some viruses. The main role of DNA molecules is the long-term storage of information. DNA is often compared to a set of blueprints, like a recipe or a code, since it contains the instructions needed to construct other components of cells, such as proteins and RNA molecules. The DNA segments that carry this genetic information are called genes, but other DNA sequences have structural purposes, or are involved in regulating the use of this genetic information.

DNA consists of two long polymers of simple units called nucleotides, with backbones made of sugars and phosphate groups joined by ester bonds. These two strands run in opposite directions to each other and are therefore anti-parallel. Attached to each sugar is one of four types of molecules called bases. It is the sequence of these four bases along the backbone that encodes information. This information is read using the genetic code, which specifies the sequence of the amino acids within proteins. The code is read by copying stretches of DNA into the related nucleic acid RNA, in a process called transcription. Within cells, DNA is organized into long structures called chromosomes. These chromosomes are duplicated before cells divide, in a process called DNA replication. Eukaryotic organisms (animals, plants, fungi, and protists) store most of their DNA inside the cell nucleus and some of their

DNA in organelles, such as mitochondria or chloroplasts. In contrast, prokaryotes (bacteria and archaea) store their DNA only in the cytoplasm. Within the chromosomes, chromatin proteins such as histones compact and organize DNA. These compact structures guide the interactions between DNA and other proteins, helping control which parts of the DNA are transcribed.

Properties

DNA is a long polymer made from repeating units called nucleotides. The DNA chain is 22 to 26 Angströms wide (2.2 to 2.6 nanometres), and one nucleotide unit is 3.3 Å (0.33 nm) long. Although each individual repeating unit is very small, DNA polymers can be very large molecules containing millions of nucleotides. For instance, the largest human chromosome, chromosome number 1, is approximately 220 million base pairs long. In living organisms, DNA does not usually exist as a single molecule, but instead as a pair of molecules that are held tightly together. These two long strands entwine like vines, in the shape of a double helix. The nucleotide repeats contain both the segment of the backbone of the molecule, which holds the chain together, and a base, which interacts with the other DNA strand in the helix. A base linked to a sugar is called a nucleoside and a base linked to a sugar and one or more phosphate groups is called a nucleotide. If multiple nucleotides are linked together, as in DNA, this polymer is called a polynucleotide.

The backbone of the DNA strand is made from alternating phosphate and sugar residues. The sugar in DNA is 2-deoxyribose, which is a pentose (five-carbon) sugar. The sugars are joined together by phosphate groups that form phosphodiester bonds between the third and fifth carbon atoms of adjacent sugar rings. These asymmetric bonds mean a strand of DNA has a direction. In a double helix the direction of the nucleotides in one strand is opposite to their direction in the other strand: the strands are *antiparallel*. The asymmetric ends of DNA strands are called the 52 (*five prime*) and 32 (*three prime*) ends, with the 5' end having a terminal phosphate group and the 3' end a terminal hydroxyl group. One major difference between DNA and RNA is the sugar, with the 2-deoxyribose in DNA being replaced by the alternative pentose sugar ribose in RNA.

The DNA double helix is stabilized by hydrogen bonds between the bases attached to the two strands. The four bases found in DNA are adenine (abbreviated A), cytosine (C), guanine (G) and thymine (T). These four bases are attached to the sugar/phosphate to form the complete nucleotide, as shown for adenosine monophosphate.

These bases are classified into two types; adenine and guanine are fused five- and six-membered heterocyclic compounds called purines, while cytosine and thymine are six-membered rings called pyrimidines. A fifth pyrimidine base, called uracil (U), usually takes the place of thymine in RNA and differs from thymine by lacking a methyl group on its ring. Uracil is not usually found in DNA, occurring only as a breakdown product of cytosine. In addition to RNA and DNA, a large number of artificial nucleic acid analogues have also been created to study the proprieties of nucleic acids, or for use in biotechnology.

Grooves

Twin helical strands form the DNA backbone. Another double helix may be found by tracing the spaces, or grooves, between the strands. These voids are adjacent to the base pairs and may provide a binding site. As the strands are not directly opposite each other, the grooves are unequally sized. One groove, the major groove, is 22 wide and the other, the minor groove, is 12 wide. The narrowness of the minor groove means that the edges of the bases are more accessible in the major groove. As a result, proteins like transcription factors that can bind to specific sequences in double-stranded DNA usually make contacts to the sides of the bases exposed in the major groove. This situation varies in unusual conformations of DNA within the cell, but the major and minor grooves are always named to reflect the differences in size that would be seen if the DNA is twisted back into the ordinary B form.

Base Pairing

Each type of base on one strand forms a bond with just one type of base on the other strand. This is called complementary base pairing. Here, purines form hydrogen bonds to pyrimidines, with A bonding only to T, and C bonding only to G. This arrangement of two nucleotides binding together across the double helix is called a base pair.

As hydrogen bonds are not covalent, they can be broken and rejoined relatively easily. The two strands of DNA in a double helix can therefore be pulled apart like a zipper, either by a mechanical force or high temperature. As a result of this complementarity, all the information in the double-stranded sequence of a DNA helix is duplicated on each strand, which is vital in DNA replication. Indeed, this reversible and specific interaction between complementary base pairs is critical for all the functions of DNA in living organisms.

The two types of base pairs form different numbers of hydrogen bonds, AT forming two hydrogen bonds, and GC forming three hydrogen bonds. DNA with high GC-content is more stable than DNA with low GC-content, but contrary to popular belief, this is not due to the extra hydrogen bond of a GC base pair but rather the contribution of stacking interactions (hydrogen bonding merely provides specificity of the pairing, not stability). As a result, it is both the percentage of GC base pairs and the overall length of a DNA double helix that determine the strength of the association between the two strands of DNA.

Long DNA helices with a high GC content have stronger-interacting strands, while short helices with high AT content have weaker-interacting strands. In biology, parts of the DNA double helix that need to separate easily, such as the TATAAT Pribnow box in some promoters, tend to have a high AT content, making the strands easier to pull apart.

In the laboratory, the strength of this interaction can be measured by finding the temperature required to break the hydrogen bonds, their melting temperature (also called T_m value). When all the base pairs in a DNA double helix melt, the strands separate and exist in solution as two entirely independent molecules. These single-stranded DNA molecules (*ssDNA*) have no single common shape, but some conformations are more stable than others.

Sense and Antisense

A DNA sequence is called "sense" if its sequence is the same as that of a messenger RNA copy that is translated into protein. The sequence on the opposite strand is called the "antisense" sequence. Both sense and antisense sequences can exist on different parts of the same strand of DNA (i.e. both strands contain both sense and antisense sequences). In both prokaryotes and eukaryotes, antisense RNA sequences are produced, but the functions of these RNAs are not entirely clear. One proposal is that antisense RNAs are involved in regulating gene expression through RNA-RNA base pairing.

A few DNA sequences in prokaryotes and eukaryotes, and more in plasmids and viruses, blur the distinction between sense and antisense strands by having overlapping genes. In these cases, some DNA sequences do double duty, encoding one protein when read along one strand, and a second protein when read in the opposite direction along the other strand. In bacteria, this overlap may be involved in the regulation of gene transcription, while in viruses, overlapping genes increase the amount of information that can be encoded within the small viral genome.

Supercoiling

DNA can be twisted like a rope in a process called DNA supercoiling. With DNA in its "relaxed" state, a strand usually circles the axis of the double helix once every 10.4 base pairs, but if the DNA is twisted the strands become more tightly or more loosely wound. If the DNA is twisted in the direction of the helix, this is positive supercoiling, and the bases are held more tightly together. If they are twisted in the opposite direction, this is negative supercoiling, and the bases come apart more easily.

In nature, most DNA has slight negative supercoiling that is introduced by enzymes called topoisomerases. These enzymes are also needed to relieve the twisting stresses introduced into DNA strands during processes such as transcription and DNA replication.

Alternate DNA Structures

DNA exists in many possible conformations that include A-DNA, B-DNA, and Z-DNA forms, although, only B-DNA and Z-DNA have been directly observed in functional organisms. The conformation that DNA adopts depends on the hydration level, DNA sequence, the amount and direction of supercoiling, chemical modifications of the bases, the type and concentration of metal ions, as well as the presence of polyamines in solution.

The first published reports of A-DNA X-ray diffraction patterns— and also B-DNA used analyses based on Patterson transforms that provided only a limited amount of structural information for oriented fibres of DNA. An alternate analysis was then proposed by Wilkins *et al.*, in 1953, for the *in vivo* B-DNA X-ray diffraction/scattering patterns of highly hydrated DNA fibres in terms of squares of Bessel functions. In the same journal, James D. Watson and Francis Crick presented their molecular modeling analysis of the DNA X-ray diffraction patterns to suggest that the structure was a double-helix.

Although the 'B-DNA form' is most common under the conditions found in cells, it is not a well-defined conformation but a family of related DNA conformations that occur at the high hydration levels present in living cells. Their corresponding X-ray diffraction and scattering patterns are characteristic of molecular paracrystals with a significant degree of disorder.

Compared to B-DNA, the A-DNA form is a wider right-handed spiral, with a shallow, wide minor groove and a narrower, deeper major groove. The A form occurs under non-physiological conditions in partially dehydrated samples of DNA, while in the cell it may be produced in hybrid pairings of DNA and RNA strands, as well as in enzyme-DNA complexes. Segments of DNA where the bases have been chemically modified by methylation may undergo a larger change in conformation and adopt the Z form. Here, the strands turn about the helical axis in a left-handed spiral, the opposite of the more common B form. These unusual structures can be recognized by specific Z-DNA binding proteins and may be involved in the regulation of transcription.

Quadruplex Structures

At the ends of the linear chromosomes are specialized regions of DNA called telomeres. The main function of these regions is to allow the cell to replicate chromosome ends using the enzyme telomerase, as the enzymes that normally replicate DNA cannot copy the extreme 32 ends of chromosomes. These specialized chromosome caps also help protect the DNA ends, and stop the DNA repair systems in the cell from treating them as damage to be corrected. In human cells, telomeres are usually lengths of single-stranded DNA containing several thousand repeats of a simple TTAGGG sequence. These guanine-rich sequences may stabilize chromosome ends by forming structures of stacked sets of four-base units, rather than the usual base pairs found in other DNA molecules. Here, four guanine bases form a flat plate and these flat four-base units then stack on top of each other, to form a stable *G-quadruplex* structure. These structures are stabilized by hydrogen bonding between the edges of the bases and chelation of a metal ion in the centre of each four-base unit. Other structures can also be formed, with the central set of four bases coming from either a single strand folded around the bases, or several different parallel strands, each contributing one base to the central structure.

In addition to these stacked structures, telomeres also form large loop structures called telomere loops, or T-loops. Here, the single-stranded DNA curls around in a long circle stabilized by telomere-binding proteins. At the very end of the T-loop, the single-stranded telomere DNA is held onto a region of double-stranded DNA by the telomere strand disrupting the double-helical DNA and base pairing to one of the two strands. This triple-stranded structure is called a displacement loop or D-loop.

Branched DNA

In DNA fraying occurs when non-complementary regions exist at the end of an otherwise complementary double-strand of DNA. However, branched DNA can occur if a third strand of DNA is introduced and contains adjoining regions able to hybridize with the frayed regions of the pre-existing double-strand. Although the simplest example of branched DNA involves only three strands of DNA, complexes involving additional strands and multiple branches are also possible.

Vibration

DNA may carry out low-frequency collective motion as observed by the Raman spectroscopy and analyzed with a quasi-continuum model.

Chemical Modifications

Base Modifications

The expression of genes is influenced by how the DNA is packaged in chromosomes, in a structure called chromatin. Base modifications can be involved in packaging, with regions that have low or no gene expression usually containing high levels of methylation of cytosine bases. For example, cytosine methylation, produces 5-methylcytosine, which is important for X-chromosome inactivation. The average level of methylation varies between organisms - the worm *Caenorhabditis elegans* lacks cytosine methylation, while vertebrates have higher levels, with up to 1% of their DNA containing 5-methylcytosine. Despite the importance of 5-methylcytosine, it can deaminate to leave a thymine base, so methylated cytosines are particularly prone to mutations. Other base modifications include adenine methylation in bacteria, the presence of 5-hydroxymethylcytosine in the brain, and the glycosylation of uracil to produce the "J-base" in kinetoplastids.

Damage

DNA can be damaged by many sorts of mutagens, which change the DNA sequence. Mutagens include oxidizing agents, alkylating agents and also high-energy electromagnetic radiation such as ultraviolet light and X-rays. The type of DNA damage produced depends on the type of mutagen. For example, UV light can damage DNA by producing thymine dimers, which are cross-links between pyrimidine bases. On the other hand, oxidants such as free radicals or hydrogen peroxide produce multiple forms of damage, including base modifications, particularly of guanosine, and double-strand breaks.

A typical human cell contains about 150,000 bases that have suffered oxidative damage. Of these oxidative lesions, the most dangerous are double-strand breaks, as these are difficult to repair and can produce point mutations, insertions and deletions from the DNA sequence, as well as chromosomal translocations.

Many mutagens fit into the space between two adjacent base pairs, this is called *intercalation*. Most intercalators are aromatic and planar molecules; examples include ethidium bromide, daunomycin, and doxorubicin. In order for an intercalator to fit between base pairs, the bases must separate, distorting the DNA strands by unwinding of the double helix. This inhibits both transcription and DNA replication, causing toxicity and mutations. As a result, DNA intercalators are often carcinogens, and benzo[*a*]pyrene diol epoxide, acridines, aflatoxin and ethidium bromide are well-known examples. Nevertheless, due to their ability to inhibit DNA transcription and replication, other similar toxins are also used in chemotherapy to inhibit rapidly growing cancer cells.

Biological Functions

DNA usually occurs as linear chromosomes in eukaryotes, and circular chromosomes in prokaryotes. The set of chromosomes in a cell makes up its genome; the human genome has approximately 3 billion base pairs of DNA arranged into 46 chromosomes. The information carried by DNA is held in the sequence of pieces of DNA called genes. Transmission of genetic information in genes is achieved via complementary base pairing.

For example, in transcription, when a cell uses the information in a gene, the DNA sequence is copied into a complementary RNA sequence through the attraction between the DNA and the correct RNA nucleotides. Usually, this RNA copy is then used to make a matching protein sequence in a process called translation, which depends on the same interaction between RNA nucleotides.

In alternative fashion, a cell may simply copy its genetic information in a process called DNA replication. The details of these functions are covered in other articles; here we focus on the interactions between DNA and other molecules that mediate the function of the genome.

Genes and Genomes

Genomic DNA is tightly and orderly packed in the process called DNA condensation to fit the small available volumes of the cell. In eukaryotes, DNA is located in the cell nucleus, as well as small amounts in mitochondria and chloroplasts. In prokaryotes, the DNA is held within an irregularly shaped body in the cytoplasm called the nucleoid. The genetic information in a genome is held within genes, and the complete set of this information in an organism is called its genotype. A gene is a unit of heredity and is a region of DNA that influences a particular characteristic in an organism. Genes contain an open reading frame that can be transcribed, as well as regulatory sequences such as promoters and enhancers, which control the transcription of the open reading frame.

In many species, only a small fraction of the total sequence of the genome encodes protein. For example, only about 1.5% of the human genome consists of protein-coding exons, with over 50% of human DNA consisting of non-coding repetitive sequences. The reasons for the presence of so much non-coding DNA in eukaryotic genomes and the extraordinary differences in genome size, or *C-value*, among species represent a long-standing puzzle known as the "C-value enigma". However, DNA sequences that do not code protein may still encode functional non-coding RNA molecules, which are involved in the regulation of gene expression.

Some non-coding DNA sequences play structural roles in chromosomes. Telomeres and centromeres typically contain few genes, but are important for the function and stability of chromosomes. An abundant form of non-coding DNA in humans are pseudogenes, which are copies of genes that have been disabled by mutation. These sequences are usually just molecular fossils, although they can occasionally serve as raw genetic material for the creation of new genes through the process of gene duplication and divergence.

Transcription and Translation

A gene is a sequence of DNA that contains genetic information and can influence the phenotype of an organism. Within a gene, the sequence of bases along a DNA strand defines a messenger RNA sequence, which then defines one or more protein sequences. The relationship between the nucleotide sequences of genes and the amino-acid sequences of proteins is determined by the rules of translation, known collectively as the genetic code. The genetic code consists of three-letter 'words' called *codons* formed from a sequence of three nucleotides (e.g. ACT, CAG, TTT).

In transcription, the codons of a gene are copied into messenger RNA by RNA polymerase. This RNA copy is then decoded by a ribosome that reads the RNA sequence by base-pairing the messenger RNA to transfer RNA, which carries amino acids. Since there are 4 bases in 3-letter combinations, there are 64 possible codons (4 combinations). These encode the twenty standard amino acids, giving most amino acids more than one possible codon. There are also three 'stop' or 'nonsense' codons signifying the end of the coding region; these are the TAA, TGA and TAG codons.

Replication

Cell division is essential for an organism to grow, but, when a cell divides, it must replicate the DNA in its genome so that the two daughter cells have the same genetic information as their parent. The double-stranded structure of DNA provides a simple mechanism for DNA replication. Here, the two strands are separated and then each strand's complementary DNA sequence is recreated by an enzyme called DNA polymerase.

This enzyme makes the complementary strand by finding the correct base through complementary base pairing, and bonding it onto the original strand. As DNA polymerases can only extend a DNA strand in a 52 to 32 direction, different mechanisms are used to copy the antiparallel strands of the double helix. In this way, the base on the old strand dictates which base appears on the new strand, and the cell ends up with a perfect copy of its DNA.

Interactions with Proteins

All the functions of DNA depend on interactions with proteins. These protein interactions can be non-specific, or the protein can bind specifically to a single DNA sequence. Enzymes can also bind to DNA and of these, the polymerases that copy the DNA base sequence in transcription and DNA replication are particularly important.

DNA-binding Proteins

Structural proteins that bind DNA are well-understood examples of non-specific DNA-protein interactions. Within chromosomes, DNA is held in complexes with structural proteins. These proteins organize the DNA into a compact structure called chromatin. In eukaryotes this structure involves DNA binding to a complex of small basic proteins called histones, while in prokaryotes multiple types of proteins are involved. The histones form a disk-shaped complex called a nucleosome, which contains two complete turns of double-stranded DNA wrapped around its surface. These non-specific interactions are formed through basic residues in the histones making ionic bonds to the acidic sugar-phosphate backbone of the DNA, and are therefore largely independent of the base sequence. Chemical modifications of these basic amino acid residues include methylation, phosphorylation and acetylation.

These chemical changes alter the strength of the interaction between the DNA and the histones, making the DNA more or less accessible to transcription factors and changing the rate of transcription. Other non-specific DNA-binding proteins in chromatin include the high-mobility group proteins, which bind to bent or distorted DNA. These proteins are important in bending arrays of nucleosomes and arranging them into the larger structures that make up chromosomes.

A distinct group of DNA-binding proteins are the DNA-binding proteins that specifically bind single-stranded DNA. In humans, replication protein A is the best-understood member of this family and is used in processes where the double helix is separated, including DNA replication, recombination and DNA repair. These binding proteins seem to stabilize single-stranded DNA and protect it from forming stem-loops or being degraded by nucleases.

In contrast, other proteins have evolved to bind to particular DNA sequences. The most intensively studied of these are the various transcription factors, which are proteins that regulate transcription. Each transcription factor binds to one particular set of DNA sequences and activates or inhibits the transcription of genes that have these sequences close to their promoters. The transcription factors do this in two ways. Firstly, they can bind the RNA polymerase responsible for transcription, either directly or through other mediator proteins; this locates the polymerase at the promoter and allows it to begin transcription. Alternatively, transcription factors can bind enzymes that modify the histones at the promoter; this will change the accessibility of the DNA template to the polymerase.

As these DNA targets can occur throughout an organism's genome, changes in the activity of one type of transcription factor can affect thousands of genes. Consequently, these proteins are often the targets of the signal transduction processes that control responses to environmental changes or cellular differentiation and development. The specificity of these transcription factors' interactions with DNA come from the proteins making multiple contacts to the edges of the DNA bases, allowing them to "read" the DNA sequence. Most of these base-interactions are made in the major groove, where the bases are most accessible.

DNA-modifying Enzymes

Nucleases and Ligases

Nucleases are enzymes that cut DNA strands by catalyzing the hydrolysis of the phosphodiester bonds. Nucleases that hydrolyse nucleotides from the ends of DNA strands are called exonucleases, while endonucleases cut within strands. The most frequently used nucleases in molecular biology are the restriction endonucleases, which cut DNA at specific sequences. For instance, the EcoRV enzyme shown to the left recognizes the 6-base sequence 52 - GAT | ATC-32 and makes a cut at the vertical line. In nature, these enzymes protect bacteria against phage infection by digesting the phage DNA when it enters the bacterial cell, acting as part of the restriction modification system. In technology, these sequence-specific nucleases are used in molecular cloning and DNA fingerprinting.

Enzymes called DNA ligases can rejoin cut or broken DNA strands. Ligases are particularly important in lagging strand DNA replication, as they join together the short segments of DNA produced at the replication fork into a complete copy of the DNA template. They are also used in DNA repair and genetic recombination.

Topoisomerases and Helicases

Topoisomerases are enzymes with both nuclease and ligase activity. These proteins change the amount of supercoiling in DNA. Some of these enzymes work by cutting the DNA helix and allowing one section to rotate, thereby reducing its level of supercoiling; the enzyme then seals the DNA break. Other types of these enzymes are capable of cutting one DNA helix and then passing a second strand of DNA through this break, before rejoining the helix. Topoisomerases are required for many processes involving DNA, such as DNA replication and transcription.

Helicases are proteins that are a type of molecular motor. They use the chemical energy in nucleoside triphosphates, predominantly ATP, to break hydrogen bonds between bases and unwind the DNA double helix into single strands. These enzymes are essential for most processes where enzymes need to access the DNA bases.

Polymerases

Polymerases are enzymes that synthesize polynucleotide chains from nucleoside triphosphates. The sequence of their products are copies of existing polynucleotide chains - which are called *templates*. These enzymes function by adding nucleotides onto the 32 hydroxyl group of the previous nucleotide in a DNA strand. As a consequence, all polymerases work in a 52 to 32 direction. In the active site of these enzymes, the incoming nucleoside triphosphate base-pairs to the template: this allows polymerases to accurately synthesize the complementary strand of their template. Polymerases are classified according to the type of template that they use.

In DNA replication, a DNA-dependent DNA polymerase makes a copy of a DNA sequence. Accuracy is vital in this process, so many of these polymerases have a proofreading activity. Here, the polymerase recognizes the occasional mistakes in the synthesis reaction by the lack of base pairing between the mismatched nucleotides. If a mismatch is detected, a 32 to 52 exonuclease activity is activated and the incorrect base removed. In most organisms, DNA polymerases function in a large complex called the replisome that contains multiple accessory subunits, such as the DNA clamp or helicases.

RNA-dependent DNA polymerases are a specialized class of polymerases that copy the sequence of an RNA strand into DNA. They include reverse transcriptase, which is a viral enzyme involved in the infection of cells by retroviruses, and telomerase, which is required for the replication of telomeres. Telomerase is an unusual polymerase because it contains its own RNA template as part of its structure.

Transcription is carried out by a DNA-dependent RNA polymerase that copies the sequence of a DNA strand into RNA. To begin transcribing a gene, the RNA polymerase binds to a sequence of DNA called a promoter and separates the DNA strands. It then copies the gene sequence into a messenger RNA transcript until it reaches a region of DNA called the terminator, where it halts and detaches from the DNA. As with human DNA-dependent DNA polymerases, RNA polymerase II, the enzyme that transcribes most of the genes in the human genome, operates as part of a large protein complex with multiple regulatory and accessory subunits.

Genetic Recombination

A DNA helix usually does not interact with other segments of DNA, and in human cells the different chromosomes even occupy separate areas in the nucleus called "chromosome territories". This physical separation of different chromosomes is important for the ability of DNA to function as a stable repository for information, as one of the few times chromosomes interact is during chromosomal crossover when they recombine. Chromosomal crossover is when two DNA helices break, swap a section and then rejoin.

Recombination allows chromosomes to exchange genetic information and produces new combinations of genes, which increases the efficiency of natural selection and can be important in the rapid evolution of new proteins. Genetic recombination can also be involved in DNA repair, particularly in the cell's response to double-strand breaks.

The most common form of chromosomal crossover is homologous recombination, where the two chromosomes involved share very similar sequences. Non-homologous recombination can be damaging to cells, as it can produce chromosomal translocations and genetic abnormalities. The recombination reaction is catalyzed by enzymes known as recombinases, such as RAD51. The first step in recombination is a double-stranded break either caused by an endonuclease or damage to the DNA. A series of steps catalyzed in part by the recombinase then leads to joining of the two helices by at least one Holliday junction, in which a segment of a single strand in each helix is annealed to the complementary strand in the other helix. The Holliday junction is a tetrahedral junction structure that can be moved along the pair of chromosomes, swapping one strand for another. The recombination reaction is then halted by cleavage of the junction and re-ligation of the released DNA.

Evolution

DNA contains the genetic information that allows all modern living things to function, grow and reproduce. However, it is unclear how long in the 4-billion-year history of life DNA has performed this function, as it has been proposed that the earliest forms of life may have used RNA as their genetic material. RNA may have acted as the central part of early cell metabolism as it can both transmit genetic information and carry out catalysis as part of ribozymes. This ancient RNA world where nucleic acid would have been used for both catalysis and genetics may have influenced the evolution of the current genetic code based on four nucleotide bases. This would occur, since the number of different bases in such an organism is a trade-off between a small number of bases increasing replication accuracy and a large number of bases increasing the catalytic efficiency of ribozymes.

However, there is no direct evidence of ancient genetic systems, as recovery of DNA from most fossils is impossible. This is because DNA will survive in the environment for less than one million years and slowly degrades into short fragments in solution. Claims for older DNA have been made, most notably a report of the isolation of a viable bacterium from a salt crystal 250 million years old, but these claims are controversial.

Uses in Technology

Genetic Engineering

Methods have been developed to purify DNA from organisms, such as phenol-chloroform extraction and manipulate it in the laboratory, such as restriction digests and the polymerase chain reaction. Modern biology and biochemistry make intensive use of these techniques in recombinant DNA technology. Recombinant DNA is a man-made DNA sequence that has been assembled from other DNA sequences. They can be transformed into organisms in the form of plasmids or in the appropriate format, by using a viral vector. The genetically modified organisms produced can be used to produce products such as recombinant proteins, used in medical research, or be grown in agriculture.

Forensics

Forensic scientists can use DNA in blood, semen, skin, saliva or hair found at a crime scene to identify a matching DNA of an individual, such as a perpetrator. This process is formally termed DNA profiling, but may also be called "genetic fingerprinting". In DNA profiling, the lengths of variable sections of repetitive DNA, such as short tandem repeats and minisatellites, are compared between people. This method is usually an extremely reliable technique for identifying a matching DNA. However, identification can be complicated if the scene is contaminated with DNA from several people. DNA profiling was developed in 1984 by British geneticist Sir Alec Jeffreys, and first used in forensic science to convict Colin Pitchfork in the 1988 Enderby murders case.

People convicted of certain types of crimes may be required to provide a sample of DNA for a database. This has helped investigators solve old cases where only a DNA sample was obtained from the scene. DNA profiling can also be used to identify victims of mass casualty incidents. On the other hand, many convicted people have been released from prison on the basis of DNA techniques, which were not available when a crime had originally been committed.

DNA Profiling

DNA profiling (also called DNA testing, DNA typing, or genetic fingerprinting) is a technique employed by forensic scientists to assist in the identification of individuals on the basis of their respective DNA profiles. DNA profiles are encrypted sets of numbers that reflect a person's DNA make-up, which can also be used as the person's identifier. DNA profiling should not be confused with full genome sequencing. It is used in, for example, parental testing and rape investigation.

Although 99.9% of human DNA sequences are the same in every person, enough of the DNA is different to distinguish one individual from another. DNA profiling uses repetitive ("repeat") sequences that are highly variable, called variable number tandem repeats (VNTR). VNTRs loci are very similar between closely related humans, but so variable that unrelated individuals are extremely unlikely to have the same VNTRs. The DNA profiling technique was first reported in 1984 by Sir Alec Jeffreys at the University of Leicester in England, and is now the basis of several national DNA databases. Dr. Jeffreys's genetic fingerprinting was made commercially available in 1987, when a chemical company, ICI, started a blood-testing centre in England.

DNA Profiling Process

The process begins with a sample of an individual's DNA (typically called a "reference sample"). The most desirable method of collecting a reference sample is the use of a buccal swab, as this reduces the possibility of contamination. When this is not available (e.g. because a court order may be needed and not obtainable) other methods may need to be used to collect a sample of blood, saliva, semen, or other appropriate fluid or tissue from personal items (e.g. toothbrush, razor, etc.) or from stored samples (e.g. banked sperm or biopsy tissue). Samples obtained from blood relatives (biological relative) can provide an indication of an individual's profile, as could human remains which had been previously profiled. A reference sample is then analyzed to create the individual's DNA profile using one of a number of techniques, discussed below. The DNA profile is then compared against another sample to determine whether there is a genetic match.

RFLP Analysis

The first methods for finding out genetics used for DNA profiling involved restriction enzyme digestion, followed by Southern blot analysis. Although polymorphisms can exist in the restriction enzyme cleavage sites, more commonly the enzymes and DNA probes were used to analyze VNTR loci. However, the Southern blot technique is laborious, and requires large amounts of undegraded sample DNA. Also, Karl Brown's original technique looked at many minisatellite loci at the same time, increasing the observed variability, but making it hard to discern individual alleles (and thereby precluding parental testing). These early techniques have been supplanted by PCR-based assays.

PCR Analysis

With the invention of the polymerase chain reaction (PCR) technique, DNA profiling took huge strides forward in both discriminating power and the ability to recover information from very small (or degraded) starting samples. PCR greatly amplifies the amounts of a specific region of DNA, using oligonucleotide primers and a thermostable DNA polymerase.

Early assays such as the HLA-DQ alpha reverse dot blot strips grew to be very popular due to their ease of use, and the speed with which a result could be obtained. However they were not as discriminating as RFLP. It was also difficult to determine a DNA profile for mixed samples, such as a vaginal swab from a sexual assault victim.

Fortunately, the PCR method is readily adaptable for analyzing VNTR loci. In the United States the FBI has standardized a set of 13 VNTR assays for DNA typing, and has organized the CODIS database for forensic identification in criminal cases. Similar assays and databases have been set up in other countries. Also, commercial kits are available that analyze single-nucleotide polymorphisms (SNPs). These kits use PCR to amplify specific regions with known variations and hybridize them to probes anchored on cards, which results in a coloured spot corresponding to the particular sequence variation.

STR Analysis

The method of DNA profiling used today is based on PCR and uses short tandem repeats (STR). This method uses highly polymorphic regions that have short repeated sequences of DNA (the most common is 4 bases repeated, but there are other lengths in use, including 3 and 5 bases). Because unrelated people almost certainly have different numbers of repeat units, STRs can be used to discriminate between unrelated individuals. These STR loci (locations on a chromosome) are targeted with sequence-specific primers and amplified using PCR. The DNA fragments that result are then separated and detected using electrophoresis. There are two common methods of separation and detection, capillary electrophoresis (CE) and gel electrophoresis.

Each STR is polymorphic, however, the number of alleles is small. Typically each STR allele will be shared by around 5 - 20% of individuals. The power of STR analysis comes from looking at multiple STR loci simultaneously. The pattern of alleles can identify an individual quite accurately. Thus STR analysis provides an excellent identification tool. The more STR regions that are tested in an individual the more discriminating the test becomes.

From country to country, different STR-based DNA-profiling systems are in use. In North America, systems which amplify the CODIS 13 core loci are almost universal, while in the UK the SGM+ system (which is compatible with The National DNA Database), is in use. Whichever system is used, many of the STR regions used are the same. These DNA-profiling systems are based on multiplex reactions, whereby many STR regions will be tested at the same time.

The true power of STR analysis is in its statistical power of discrimination. Because the 13 loci that are currently used for discrimination in CODIS are independently assorted (having a certain number of repeats at one locus doesn't change the likelihood of having any number of repeats at any other locus), the product rule for probabilities can be applied. This means that if someone has the DNA type of ABC, where the three loci were independent, we can say that the probability of having that DNA type is the probability of having type A times the probability of having type B times the probability of having type C. This has resulted in the ability to generate match probabilities of 1 in a quintillion (1 with 18 zeros after it) or more.

However, DNA database searches showed much more frequent than expected false DNA matches including one perfect 13 locus match out of only 30,000 DNA samples in Maryland in January 2007.

Moreover, since there are about 12 million monozygotic twins on Earth, that theoretical probability is useless. For example, the actual probability that 2 random people have the same DNA depends on whether there were twins or triplets (etc.) in the family, and the number of loci used in the test. Where twins are common, the probability of matching the DNA is 22 in 1000, or about 2.2 in 100 will have matching DNA.

In practice, the risk of contaminated-matching is much greater than matching a distant relative, such as a sample being contaminated from nearby objects, or from left-over cells transferred from a prior test. Logically, the risk is greater for matching the most common person in the samples: everything collected from, or in contact with, a victim is a major source of contamination for any other samples brought into a lab. For that reason, multiple control-samples are typically tested, to ensure that they stayed clean, when prepared during the same period as the actual test samples. Unexpected matches (or variations) in several control-samples indicates a high probability of contamination for the actual test samples. In a relationship test, the full DNA profiles should differ (except for twins), to prove that a person wasn't actually matched as being related to their own DNA in another sample.

AmpFLP

Another technique, AmpFLP, or amplified fragment length polymorphism was also put into practice during the early 1990s. This technique was also faster than RFLP analysis and used PCR to amplify DNA samples. It relied on variable number tandem repeat (VNTR) polymorphisms to distinguish various alleles, which were separated on a polyacrylamide gel using an allelic ladder (as opposed to a molecular weight ladder). Bands could be visualized by silver staining the gel. One popular locus for fingerprinting was the D1S80 locus.

As with all PCR based methods, highly degraded DNA or very small amounts of DNA may cause allelic dropout (causing a mistake in thinking a heterozygote is a homozygote) or other stochastic effects. In addition, because the analysis is done on a gel, very high number repeats may bunch together at the top of the gel, making it difficult to resolve. AmpFLP analysis can be highly automated, and allows for easy creation of phylogenetic trees based on comparing individual samples of DNA. Due to its relatively low cost and ease of set-up and operation, AmpFLP remains popular in lower income countries.

DNA Family Relationship Analysis

Using PCR technology, DNA analysis is widely applied to determine genetic family relationships such as paternity, maternity, siblingship and other kinships.

During conception, the father's sperm cell and the mother's egg cell, each containing half the amount of DNA found in other body cells, meet and fuse to form a fertilized egg, called a zygote. The zygote contains a complete set of DNA molecules, a unique combination of DNA from both parents. This zygote divides and multiplies into an embryo and later, a full human being.

At each stage of development, all the cells forming the body contain the same DNA—half from the father and half from the mother. This fact allows the relationship testing to use all types of all samples including loose cells from the cheeks collected using buccal swabs, blood or other types of samples. While a lot of DNA contains information for a certain function, there is some called junk DNA, which is currently used for human identification. At some special locations (called loci) in the junk DNA, predictable inheritance patterns were found to be useful in determining biological relationships. These locations contain specific DNA markers that DNA scientists use to identify individuals. In a routine DNA paternity test, the markers used are Short Tandem Repeats (STRs), short pieces of DNA that occur in highly differential repeat patterns among individuals.

Each person's DNA contains two copies of these markers—one copy inherited from the father and one from the mother. Within a population, the markers at each person's DNA location could differ in length and sometimes sequence, depending on the markers inherited from the parents.

The combination of marker sizes found in each person makes up his/her unique genetic profile. When determining the relationship between two individuals, their genetic profiles are compared to see if they share the same inheritance patterns at a statistically conclusive rate.

For example, the following sample report from this commercial DNA paternity testing laboratory Universal Genetics signifies how relatedness between parents and child is identified on those special markers:

DNA Marker	*Mother*	*Child*	*Alleged father*
D21S11	28, 30	28, 31	29, 31
D7S820	9, 10	10, 11	11, 12
TH01	14, 15	14, 16	15, 16
D13S317	7, 8	7, 9	8, 9
D19S433	14, 16.2	14, 15	15, 17

The partial results indicate that the child and the alleged father's DNA match among these five markers. The complete test results show this correlation on 16 markers between the child and the tested man to draw a conclusion of whether or not the man is the biological father.

Scientifically, each marker is assigned with a Paternity Index (PI), which is a statistical measure of how powerfully a match at a particular marker indicates paternity. The PI of each marker is multiplied with each other to generate the Combined Paternity Index (CPI), which indicates the overall probability of an individual being the biological father of the tested child relative to any random man from the entire population of the same race. The CPI is then converted into a Probability of Paternity showing the degree of relatedness between the alleged father and child.

The DNA test report in other family relationship tests, such as grandparentage and siblingship tests, is similar to a paternity test report. Instead of the Combined Paternity Index, a different value, such as a Siblingship

Index, is reported. The report shows the genetic profiles of each tested person. If there are markers shared among the tested individuals, the probability of biological relationship is calculated to determine how likely the tested individuals share the same markers due to a blood relationship.

Y-chromosome Analysis

Recent innovations have included the creation of primers targeting polymorphic regions on the Y-chromosome (Y-STR), which allows resolution of a mixed DNA sample from a male and female and/or cases in which a differential extraction is not possible. Y-chromosomes are paternally inherited, so Y-STR analysis can help in the identification of paternally related males. Y-STR analysis was performed in the Sally Hemings controversy to determine if Thomas Jefferson had sired a son with one of his slaves.

Mitochondrial Analysis

For highly degraded samples, it is sometimes impossible to get a complete profile of the 13 CODIS STRs. In these situations, mitochondrial DNA (mtDNA) is sometimes typed due to there being many copies of mtDNA in a cell, while there may only be 1-2 copies of the nuclear DNA.

Forensic scientists amplify the HV1 and HV2 regions of the mtDNA, then sequence each region and compare single-nucleotide differences to a reference. Because mtDNA is maternally inherited, directly linked maternal relatives can be used as match references, such as one's maternal grandmother's daughter's son.

A difference of two or more nucleotides is generally considered to be an exclusion. Heteroplasmy and poly-C differences may throw off straight sequence comparisons, so some expertise on the part of the analyst is required. mtDNA is useful in determining clear identities, such as those of missing people when a maternally linked relative can be found. mtDNA testing was used in determining that Anna Anderson was not the Russian princess she had claimed to be, Anastasia Romanov.

mtDNA can be obtained from such material as hair shafts and old bones/ teeth..

DNA Databases

There are now several DNA databases in existence around the world. Some are private, but most of the largest databases are government controlled. The United States maintains the largest DNA database, with the Combined DNA Index System, holding over 5 million records as of 2007.

The United Kingdom maintains the National DNA Database (NDNAD), which is of similar size, despite the UK's smaller population. The size of this database, and its rate of growth, is giving concern to civil liberties groups in the UK, where police have wide-ranging powers to take samples and retain them even in the event of acquittal. The U.S. Patriot Act of the United States provides a means for the U.S. government to get DNA samples from other countries if they are either a division of, or head office of, a company operating in the U.S. Under the act, the American offices of the company can't divulge to their subsidiaries/offices in other countries the reasons that these DNA samples are sought or by whom.

When a match is made from a National DNA Databank to link a crime scene to an offender who has provided a DNA Sample to a databank that link is often referred to as a *cold hit.* A cold hit is of value in referring the police agency to a specific suspect but is of less evidential value than a DNA match made from outside the DNA Databank.

Considerations when Evaluating DNA Evidence

In the early days of the use of genetic fingerprinting as criminal evidence, juries were often swayed by spurious statistical arguments by defence lawyers along these lines: given a match that had a 1 in 5 million probability of occurring by chance, the lawyer would argue that this meant that in a country of say 60 million people there were 12 people who would also match the profile. This was then translated to a 1 in 12 chance of the suspect being the guilty one. This argument is not sound unless the suspect was drawn at random from the population of the country.

In fact, a jury should consider how likely it is that an individual matching the genetic profile would also have been a suspect in the case for other reasons. Another spurious statistical argument is based on the false assumption that a 1 in 5 million probability of a match automatically translates into a 1 in 5 million probability of innocence and is known as the prosecutor's fallacy.

When using RFLP, the theoretical risk of a coincidental match is 1 in 100 billion (100,000,000,000), although the practical risk is actually 1 in 1000 because monozygotic twins are 0.2% of the human population. Moreover, the rate of laboratory error is almost certainly higher than this, and often actual laboratory procedures do not reflect the theory under which the coincidence probabilities were computed.

For example, the coincidence probabilities may be calculated based on the probabilities that markers in two samples have bands in *precisely* the same location, but a laboratory worker may conclude that similar—but not precisely identical—band patterns result from identical genetic samples with some imperfection in the agarose gel. However, in this case, the laboratory worker increases the coincidence risk by expanding the criteria for declaring a match.

Recent studies have quoted relatively high error rates which may be cause for concern. In the early days of genetic fingerprinting, the necessary population data to accurately compute a match probability was sometimes unavailable. Between 1992 and 1996, arbitrary low ceilings were controversially put on match probabilities used in RFLP analysis rather than the higher theoretically computed ones. Today, RFLP has become widely disused due to the advent of more discriminating, sensitive and easier technologies.

STRs do not suffer from such subjectivity and provide similar power of discrimination (1 in 10^13 for unrelated individuals if using a full SGM+ profile) It should be noted that figures of this magnitude are not considered to be statistically supportable by scientists in the UK, for unrelated individuals with full matching DNA profiles a match probability of 1 in a billion is considered statistically supportable (Since 1998 the DNA profiling system supported by The National DNA Database in the UK is the SGM+ DNA profiling system which includes 10 STR regions and a sex indicating test. However, with any DNA technique, the cautious juror should not convict on genetic fingerprint evidence alone if other factors raise doubt. Contamination with other evidence (secondary transfer) is a key source of incorrect DNA profiles and raising doubts as to whether a sample has been adulterated is a favorite defence technique. More rarely, chimerism is one such instance where the lack of a genetic match may unfairly exclude a suspect.

Evidence of Genetic Relationship

It's also possible to use DNA profiling as evidence of genetic relationship, but testing that shows no relationship isn't absolutely certain. While almost all individuals have a single and distinct set of genes, rare individuals, known as "chimeras", have at least two different sets of genes. There have been several cases of DNA profiling that falsely "proved" that a mother was unrelated to her children.

Fake DNA Evidence

The value of DNA evidence has to be seen in light of recent cases where criminals planted fake DNA samples at crime scenes. In one case, a criminal even planted fake DNA evidence in his own body: Dr. John Schneeberger raped one of his sedated patients in 1992 and left semen on her underwear. Police drew what they believed to be Schneeberger's blood and compared its DNA against the crime scene semen DNA on three occasions, never showing a match. It turned out that he had surgically inserted a Penrose drain into his arm and filled it with foreign blood and anticoagulants.

In a study conducted by the life science company Nucleix and published in the journal Forensic Science International, scientists found that an In vitro synthesized sample of DNA matching any desired genetic profile can be constructed using standard molecular biology techniques without obtaining any actual tissue from that person.

DNA Evidence as Evidence in Criminal Trials

Familial Searching

Familial searching is the use of family members' DNA to identify a closely related suspect in jurisdictions where large DNA databases exist, but no exact match has been found. The first successful use of the practice was in a UK case where a man was convicted of manslaughter when he threw a brick stained with his own blood into a moving car.

Police could not get an exact match to the UK's DNA database because the man had no criminal convictions, but police implicated him using a close relative's DNA. The technique was used to catch a Los Angeles serial killer known as the "Grim Sleeper" in 2010. However, critics have challenged the technology as "deeply antithetical to American values" and likely to lead to excess investigation of racial and ethnic minorities.

Surreptitious DNA Collecting

Police forces may collect DNA samples without the suspects' knowledge, and use it as evidence. Legality of this mode of proceeding has been questioned in Australia. In the United States, it has been accepted, courts often claiming that there was no expectation of privacy, citing California v. Greenwood (1985), during which the Supreme Court held that the Fourth Amendment does not prohibit the warrantless search and seizure of garbage left for collection outside the curtilage of a home. Critics of this practice underline the fact that this analogy ignores that "most people have no idea that they risk surrendering their genetic identity to the police by, for instance, failing to destroy a used coffee cup.

Moreover, even if they do realize it, there is no way to avoid abandoning one's DNA in public." In the UK, the Human Tissue Act 2004 prohibited private individuals from covertly collecting biological samples (hair, fingernails, etc.) for DNA analysis, but excluded medical and criminal investigations from the offense.

England and Wales

Evidence from an expert who has compared DNA samples must be accompanied by evidence as to the sources of the samples and the procedures for obtaining the DNA profiles. The judge must ensure that the jury must understand the significance of DNA matches and mismatches in the profiles. The judge must also ensure that the jury does not confuse the 'match probability' (the probability that a person that is chosen at random has a matching DNA profile to the sample from the scene) with the 'likelihood ratio' (the probability that a person with matching DNA committed the crime). In *R v. Doheny*, EWCA Crim 728. Phillips LJ gave this example of a summing up, which should be carefully tailored to the particular facts in each case:

Members of the Jury, if you accept the scientific evidence called by the Crown, this indicates that there are probably only four or five white males in the United Kingdom from whom that semen stain could have come. The Defendant is one of them. If that is the position, the decision you have to reach, on all the evidence, is whether you are sure that it was the Defendant who left that stain or whether it is possible that it was one of that other small group of men who share the same DNA characteristics.

Juries should weigh up conflicting and corroborative evidence, using their own common sense and not by using mathematical formulae, such as Bayes' theorem, so as to avoid "confusion, misunderstanding and misjudgment".

Presentation and Evaluation of Evidence of Partial or Incomplete DNA Profiles

R v Bates (2006) EWCA Crim 1395 Moore-Bick LJ said: "We can see no reason why partial profile DNA evidence should not be admissible provided that the jury are made aware of its inherent limitations and are given a sufficient explanation to enable them to evaluate it.

There may be cases where the match probability in relation to all the samples tested is so great that the judge would consider its probative value to be minimal and decide to exclude the evidence in the exercise of his discretion, but this gives rise to no new question of principle and can be left for decision on a case by case basis.

However, the fact that there exists in the case of all partial profile evidence the possibility that a "missing" allele might exculpate the accused altogether does not provide sufficient grounds for rejecting such evidence. In many there is a possibility (at least in theory) that evidence exists which would assist the accused and perhaps even exculpate him altogether, but that does not provide grounds for excluding relevant evidence that is available and otherwise admissible, though it does make it important to ensure that the jury are given sufficient information to enable them to evaluate that evidence properly".

DNA Testing in the US

There are state laws on DNA profiling in all 50 states of the United States. Detailed information on database laws in each state can be found at the National Conference of State Legislatures website.

Development of Artificial DNA

In August 2009, scientists in Israel stunned the forensic sciences and raised serious questions concerning the use of DNA by law enforcement as the ultimate method of identification. In a paper published in the journal *Forensic Science International: Genetics*, the Israeli researchers demonstrated that it is possible to manufacture DNA in a laboratory, and thus falsify DNA evidence.

The scientists fabricated saliva and blood samples, which originally contained DNA from a person other than the ostensible donor of the blood and saliva.

The researchers also showed that, using a DNA database, it is possible to take information from a profile and manufacture DNA to match it, and that this can be done without access to any actual DNA from the person whose DNA they are duplicating. The synthetic DNA oligos required for the procedure are common in molecular laboratories.

Dr. Daniel Frumkin, lead author on the paper, was quoted in *The New York Times* as saying, "You can just engineer a crime scene... any biology undergraduate could perform this."

Dr. Frumkin perfected a test that can forensically differentiate real DNA samples from fake ones. His test detects epigenetic modifications, in particular, DNA methylation. Seventy percent of the DNA in any human genome is methylated, meaning it contains methyl group modifications within a CpG dinucleotide context. Methylation at the promoter region is associated with gene silencing. The synthetic DNA lacks this epigenetic modification, which allows the test to distinguish manufactured DNA from original, genuine, DNA.

It is unknown how many, if any, police departments currently use the test, which appears to be a serious issue in light of Frumkin's claim that the DNA manufacturing procedure is within the grasp of any undergraduate biology student. No police lab has publicly announced that it is using the new test to verify DNA results, while FSI Genetics says that any forensic laboratory doing DNA identification should adopt this test to authenticate its results as "real" DNA.

Bioinformatics

Bioinformatics involves the manipulation, searching, and data mining of DNA sequence data. The development of techniques to store and search DNA sequences have led to widely applied advances in computer science, especially string searching algorithms, machine learning and database theory. String searching or matching algorithms, which find an occurrence of a sequence of letters inside a larger sequence of letters, were developed to search for specific sequences of nucleotides. In other applications such as text editors, even simple algorithms for this problem usually suffice, but DNA sequences cause these algorithms to exhibit near-worst-case behaviour due to their small number of distinct characters.

The related problem of sequence alignment aims to identify homologous sequences and locate the specific mutations that make them distinct. These techniques, especially multiple sequence alignment, are used in studying phylogenetic relationships and protein function. Data sets representing entire genomes' worth of DNA sequences, such as those produced by the Human

Genome Project, are difficult to use without annotations, which label the locations of genes and regulatory elements on each chromosome. Regions of DNA sequence that have the characteristic patterns associated with protein- or RNA-coding genes can be identified by gene finding algorithms, which allow researchers to predict the presence of particular gene products in an organism even before they have been isolated experimentally.

DNA Nanotechnology

DNA nanotechnology uses the unique molecular recognition properties of DNA and other nucleic acids to create self-assembling branched DNA complexes with useful properties. DNA is thus used as a structural material rather than as a carrier of biological information.

This has led to the creation of two-dimensional periodic lattices (both tile-based as well as using the "DNA origami" method) as well as three-dimensional structures in the shapes of polyhedra. Nanomechanical devices and algorithmic self-assembly have also been demonstrated, and these DNA structures have been used to template the arrangement of other molecules such as gold nanoparticles and streptavidin proteins.

History and Anthropology

Because DNA collects mutations over time, which are then inherited, it contains historical information, and, by comparing DNA sequences, geneticists can infer the evolutionary history of organisms, their phylogeny. This field of phylogenetics is a powerful tool in evolutionary biology. If DNA sequences within a species are compared, population geneticists can learn the history of particular populations. This can be used in studies ranging from ecological genetics to anthropology; For example, DNA evidence is being used to try to identify the Ten Lost Tribes of Israel.

DNA has also been used to look at modern family relationships, such as establishing family relationships between the descendants of Sally Hemings and Thomas Jefferson. This usage is closely related to the use of DNA in criminal investigations detailed above. Indeed, some criminal investigations have been solved when DNA from crime scenes has matched relatives of the guilty individual.

History of DNA Research

DNA was first isolated by the Swiss physician Friedrich Miescher who, in 1869, discovered a microscopic substance in the pus of discarded surgical bandages. As it resided in the nuclei of cells, he called it "nuclein". In 1919, Phoebus Levene identified the base, sugar and phosphate nucleotide unit. Levene suggested that DNA consisted of a string of nucleotide units linked together through the phosphate groups. However, Levene thought the chain was short and the bases repeated in a fixed order. In 1937 William Astbury produced the first X-ray diffraction patterns that showed that DNA had a regular structure.

In 1928, Frederick Griffith discovered that traits of the "smooth" form of the *Pneumococcus* could be transferred to the "rough" form of the same bacteria by mixing killed "smooth" bacteria with the live "rough" form. This system provided the first clear suggestion that DNA carries genetic information—the Avery-MacLeod-McCarty experiment—when Oswald Avery, along with coworkers Colin MacLeod and Maclyn McCarty, identified DNA as the transforming principle in 1943. DNA's role in heredity was confirmed in 1952, when Alfred Hershey and Martha Chase in the Hershey-Chase experiment showed that DNA is the genetic material of the T2 phage.

In 1953, James D. Watson and Francis Crick suggested what is now accepted as the first correct double-helix model of DNA structure in the journal *Nature*. Their double-helix, molecular model of DNA was then based on a single X-ray diffraction image (labeled as "Photo 51") taken by Rosalind Franklin and Raymond Gosling in May 1952, as well as the information that the DNA bases are paired — also obtained through private communications from Erwin Chargaff in the previous years. Chargaff's rules played a very important role in establishing double-helix configurations for B-DNA as well as A-DNA.

Experimental evidence supporting the Watson and Crick model were published in a series of five articles in the same issue of *Nature*. Of these, Franklin and Gosling's paper was the first publication of their own X-ray diffraction data and original analysis method that partially supported the Watson and Crick model; this issue also contained an article on DNA structure by Maurice Wilkins and two of his colleagues, whose analysis and *in vivo* B-DNA X-ray patterns also supported the presence *in vivo* of the double-helical DNA configurations as proposed by Crick and Watson for their double-helix molecular model of DNA in the previous two pages of *Nature*. In 1962, after Franklin's death, Watson, Crick, and Wilkins jointly received the Nobel Prize in Physiology or Medicine. However, Nobel rules of the time allowed only living recipients, but a vigorous debate continues on who should receive credit for the discovery.

In an influential presentation in 1957, Crick laid out the "Central Dogma" of molecular biology, which foretold the relationship between DNA, RNA, and proteins, and articulated the "adaptor hypothesis". Final confirmation of the replication mechanism that was implied by the double-helical structure followed in 1958 through the Meselson-Stahl experiment. Further work by Crick and coworkers showed that the genetic code was based on non-overlapping triplets of bases, called codons, allowing Har Gobind Khorana, Robert W. Holley and Marshall Warren Nirenberg to decipher the genetic code. These findings represent the birth of molecular biology.

Protein Synthesis

Protein synthesis is the process in which cells build proteins. The term is sometimes used to refer only to protein translation but more often it refers to a multi-step process, beginning with amino acid synthesis and transcription of nuclear DNA into messenger RNA, which is then used as input to translation.

The cistron DNA is transcribed into a variety of RNA intermediates. The last version is used as a template in synthesis of a polypeptide chain. Proteins can often be synthesized directly from genes by translating mRNA.

When a protein needs to be available on short notice or in large quantities, a protein precursor is produced. A proprotein is an inactive protein containing one or more inhibitory peptides that can be activated when the inhibitory sequence is removed by proteolysis during posttranslational modification.

A preprotein is a form that contains a signal sequence (an N-terminal signal peptide) that specifies its insertion into or through membranes; i.e., targets them for secretion. The signal peptide is cleaved off in the endoplasmic reticulum. Preproproteins have both sequences (inhibitory and signal) still present.

For synthesis of protein, a succession of tRNA molecules charged with appropriate amino acids have to be brought together with an mRNA molecule and matched up by base-pairing through their anti-codons with each of its successive codons.

The amino acids then have to be linked together to extend the growing protein chain, and the tRNAs, relieved of their burdens, have to be released. This whole complex of processes is carried out by a giant multimolecular machine, the ribosome, formed of two main chains of RNA, called ribosomal RNA (rRNA), and more than 50 different proteins.

This molecular juggernaut latches onto the end of an mRNA molecule and then trundles along it, capturing loaded tRNA molecules and stitching together the amino acids they carry to form a new protein chain. Protein biosynthesis, although very similar, is different for prokaryotes and eukaryotes.

Amino Acid Synthesis

Amino acid synthesis is the set of biochemical processes (metabolic pathways) by which the various amino acids are produced from other compounds. The substrates for these processes are various compounds in the organism's diet or growth media. Not all organisms are able to synthesise all amino acids, for example humans are only able to synthesise 12 of the 20 standard amino acids. A fundamental problem for biological systems is to obtain nitrogen in an easily usable form. This problem is solved by certain microorganisms capable of reducing the inert Na"N molecule (nitrogen gas) to two molecules of ammonia in one of the most remarkable reactions in biochemistry.

Nitrogen in the form of ammonia is the source of nitrogen for all the amino acids. The carbon backbones come from the glycolytic pathway, the pentose phosphate pathway, or the citric acid cycle.

In amino acid production, one encounters an important problem in biosynthesis— namely, stereochemical control. Because all amino acids except glycine are chiral, biosynthetic pathways must generate the correct isomer with high fidelity.

In each of the 19 pathways for the generation of chiral amino acids, the stereochemistry at the á-carbon atom is established by a transamination reaction that involves pyridoxal phosphate. Almost all the transaminases that catalyze these reactions descend from a common ancestor, illustrating once again that effective solutions to biochemical problems are retained throughout evolution.

Biosynthetic pathways are often highly regulated such that building blocks are synthesized only when supplies are low. Very often, a high concentration of the final product of a pathway inhibits the activity of enzymes that function early in the pathway. Often present are allosteric enzymes capable of sensing and responding to concentrations of regulatory species. These enzymes are similar in functional properties to aspartate transcarbamoylase and its regulators. Feedback and allosteric mechanisms ensure that all twenty amino acids are maintained in sufficient amounts for protein synthesis and other processes.

Amino acids are synthesized from α-ketoacids, and later transaminated from another aminoacid, usually Glutamate. The enzyme involved in this reaction is an aminotransferase.

Glutamate itself is formed by amination of α-ketoglutarate:

Nitrogen Fixation: Microorganisms Use ATP and a Powerful Reductant to Reduce Atmospheric Nitrogen to Ammonia

Microorganisms use ATP and reduced ferredoxin, a powerful reductant, to reduce N2 to NH3. An iron-molybdenum cluster in nitrogenase deftly catalyzes the fixation of N2, a very inert molecule. Higher organisms consume the fixed nitrogen to synthesize amino acids, nucleotides, and other nitrogen-containing biomolecules. The major points of entry of NH4+ into metabolism are glutamine or glutamate.

Amino Acids are Made from Intermediates of the Citric Acid Cycle and Other Major Pathways

Of the basic set of 20 amino acids (not counting selenocysteine) there are 8 that human beings cannot synthesize. In addition, the amino acids arginine, cysteine, glycine, glutamine, histidine, proline, serine and tyrosine are considered conditionally essential, meaning they are not normally required in the diet, but must be supplied exogenously to specific populations that do not synthesize it in adequate amounts. (For example, enough arginine is synthesized by the urea cycle to meet the needs of an adult but perhaps not those of a growing child.)

Amino acids that need to be obtained from the diet are called essential amino acids. Nonessential amino acids are produced in the body. The pathways for the synthesis of nonessential amino acids are quite simple. Glutamate dehydrogenase catalyzes the reductive amination of α-ketoglutarate to glutamate.

A transamination reaction takes place in the synthesis of most amino acids. At this step, the chirality of the amino acid is established. Alanine and aspartate are synthesized by the transamination of pyruvate and oxaloacetate, respectively. Glutamine is synthesized from NH4+ and glutamate, and asparagine is synthesized similarly. Proline and arginine are derived from glutamate.

Serine, formed from 3-phosphoglycerate, is the precursor of glycine and cysteine. Tyrosine is synthesized by the hydroxylation of phenylalanine, an essential amino acid. The pathways for the biosynthesis of essential amino acids are much more complex than those for the nonessential ones.

Tetrahydrofolate, a carrier of activated one-carbon units, plays an important role in the metabolism of amino acids and nucleotides. This coenzyme carries one-carbon units at three oxidation states, which are interconvertible: most reduced—methyl; intermediate—methylene; and most oxidized—formyl, formimino, and methenyl.

The major donor of activated methyl groups is S-adenosylmethionine, which is synthesized by the transfer of an adenosyl group from ATP to the sulfur atom of methionine. S-Adenosylhomocysteine is formed when the activated methyl group is transferred to an acceptor. It is hydrolyzed to adenosine and homocysteine, the latter of which is then methylated to methionine to complete the activated methyl cycle. The A Cortisol inhibites protein synthesis.

Amino Acid Biosynthesis is Regulated by Feedback Inhibition

Most of the pathways of amino acid biosynthesis are regulated by feedback inhibition, in which the committed step is allosterically inhibited by the final product. Branched pathways require extensive interaction among the branches that includes both negative and positive regulation. The regulation of glutamine synthetase from E. coli is a striking demonstration of cumulative feedback inhibition and of control by a cascade of reversible covalent modifications.

Amino Acids are Precursors of Many Biomolecules

Amino acids are precursors of a variety of biomolecules. Glutathione serves as a sulfhydryl buffer and detoxifying agent. Glutathione peroxidase, a selenoenzyme, catalyzes the reduction of hydrogen peroxide and organic peroxides by glutathione. Nitric oxide, a short-lived messenger, is formed from arginine.

Porphyrins are synthesized from glycine and succinyl CoA, which condense to give β-aminolevulinate. Two molecules of this intermediate become linked to form porphobilinogen. Four molecules of porphobilinogen combine to form a linear tetrapyrrole, which cyclizes to uroporphyrinogen III. Oxidation and side-chain modifications lead to the synthesis of protoporphyrin IX, which acquires an iron atom to form heme.

Bibliography

Ayala, F. J.: *The Genetic Structure of Populations*, W.H. Freeman & Co., San Francisco, California, 1977.

Barnaby, W.: *Plague Makers: The Secret World of Biological Warfare*, Vision Paperbacks, London, 1999.

Berg, Paul, and Singer, Maxine: *Dealing with Genes: The Language of Heredity. Mill Valley*, University Science Books, CA, 1992.

Berry, R. J.: *Inheritance and Natural History*: Collins, London, 1977.

Bradley, W.G.: *Neurology in Clinical Practice*, Butterworth Heinemann, Boston, 2000.

Crick, F.: *Life Itself: Its Origin and Nature*, W.W. Norton, New York, 1982.

Curtis, M.: *The Geometry of DNA: A Structural Revision*, Blue Gallery, London, 1996.

Dawkins, R.: *The Selfish Gene*, Oxford University Press, New York, 1976.

Dunbar, Robert E.: *Heredity*, Franklin Watts Publisher, New York, 1978.

Fisher, R. A.: *The Genetical Theory of Natural Selection*, Clarendon Press, Oxford, 1930.

Glover, D. M., and Hames, B. D.: *Genes and Embryos*, Oxford University Press, New York, 1989.

Goldman, L.; Ausiello, D. A.: *Cecil Textbook of Medicine*, Saunders, Philadelphia, 2004.

Hamerton, J. L.: *Human Cytogenetics*, Academic Press, New York, 1971.

Hanley, R.: *Is Data Human? The Metaphysics of Star Trek*, Boxtree, London, 1998.

Jonoska, N.: *Self-Assembling DNA Graphs, DNA-Based Computers VIII*, Springer-Verlag, Berlin, 2003.

Kelves, D.: *In the Name of Eugenics: Genetics and the Uses of Human Heredity*. Harmondsworth: Penguin.

Klein, Aaron E.: *Threads of Life: Genetics from Aristotle to DNA*, The Natural History Press, Garden City, New York, 1955.

Lewin B.: *Genes VII*, Oxford University Press Inc., New York, USA, 2000.

Lewontin, R.C.: *The Doctrine of DNA: Biology as Ideology*, Penguin, Harmondsworth, 1993.

Lyon, Jeff and Gorner, Peter. Altered Fates: *Gene Therapy and the Retooling of Human Life*, W. W. Norton and Company, New York, 1995.

Margulis, L.: *Symbiosis in Cell Evolution,* W.H. Freeman, San Francisco, 1981.

Nelkin, D. and M.S. Lindee: *The DNA Mystique: The Gene as a Cultural Icon*, W.H. Freeman, New York, 1995.

Rietman, Ed.: *Molecular Engineering of Nanosystems,* Springer, New York, 2001.

Stebbins, G. L.: *Darwin to DNA, Molecules to Humanity*, W. H. Freeman, San Francisco, 1982.

Sturtevant, Alfred: *History of Genetics*, Harper and Row, New York, 1965.

Suzuki, D., and Knudtson, P.: *Genethics: The Clash Between the New Genetics and Human Values*, Harvard University Press, Cambridge, Mass, 1989.

Watson, J. D.: *The Double Helix*, Antheneum, New York, 1968.

Williams, J.G. and R.K. Patient: *Genetic Engineering*, IRL Press, Oxford, England, 1988.

Winchester, A. M.: *Heredity, Evolution and Humankind*, West Publishing Co., St. Paul, Minn., 1976.

Winston, R.: *The Future of Genetic Manipulation*, Phoenix, London, 1997.

Wright, L.: *Twins: Genes, Environment and the Mystery of Human Identity*, Weidenfeld & Nicolson, London, 1997.

Index

N

P

Q

R

S

T

U

❑❑❑